Instrumentation

Transducers, Experimentation, & Applications

by
Roger W. Prewitt, Associate Professor
and
Stephen W. Fardo, Associate Professor

Department of Industrial Education and Technology
College of Applied Arts and Technology
Eastern Kentucky University

Howard W. Sams & Co., Inc.
4300 WEST 62ND ST. INDIANAPOLIS, INDIANA 46268 USA

Preface

This laboratory oriented activities manual can be used by vocational/technical or college-level instrumentation programs. The unique aspect of this manual is that it does not deal with the use of specific instruments. However, completion of the activities in this manual will provide the student with an in-depth understanding of instrumentation and measurement. The student will learn *how* measurement systems operate, not just how to operate a particular instrument. The cost of the lab equipment required to complete these activities is very low. No massive equipment is needed.

The approach taken in this manual is to acquaint the student initially with instrumentation and measurement systems. The basics of measurement, such as units of measurement, scientific notation, and metric measurement, are presented in the first unit of activities. Basic meter design and measurement circuits are then investigated. These include current, voltage, and resistance measurements, bridge circuits, oscilloscope measurements, and numerical readout devices.

Unit 2 deals with thermoelectric transducers used for instrumentation systems, such as thermocouples, thermistors, and rtd's.

Unit 3 investigates photoelectric transducers, such as photoconductive, photoemissive, and photovoltaic devices, that are used in instrumentation systems.

Unit 4 is designed to explore miscellaneous types of transducers, including potentiometric, capacitive, inductive, and resistive transducers.

Unit 5 provides some practical applications of instrumentation circuits for measuring physical quantities. These include the measurement of humidity, pressure, gas flow, liquid flow, and pH.

This manual presents a basic approach to instrumentation. Any institution which has electricity/electronics programs could build an instrumentation course around it. The basic equipment and materials, such as power supplies, meters, resistors, capacitors, and potentiometers, would most likely be available in any laboratory. The remaining materials necessary to complete these activities are primarily small components, such as photoelectric and thermoelectric devices, which are used as instrumentation transducers. The cost would be exceptionally low for adding a one-semester course in instrumentation. Specialized instrumentation techniques, such as calibration, are beyond the scope of this manual.

The authors would like to thank their wives, Mary Ann and Helen, for their cooperation and patience during the preparation of this manuscript.

ROGER W. PREWITT
STEPHEN W. FARDO

Laboratory Equipment and Components

The following list of equipment and components is necessary for the successful completion of the activities included in this laboratory manual.

Resistors (1 W unless specified)

100 Ω
150 Ω
220 Ω, 2 W
270 Ω
365 Ω
390 Ω
470 Ω (8)
560 Ω
650 Ω
680 Ω
1 kΩ
1.2 kΩ, 20 W
2 kΩ
2.7 kΩ
4.7 kΩ
5.6 kΩ
6.8 kΩ
10 kΩ, 2 W
12 kΩ
15 kΩ
20 kΩ
25 kΩ
33 kΩ
120 kΩ
100 kΩ (2)
200 kΩ (2)
1 MΩ
5.6 MΩ

Potentiometers (1 W unless specified)

200 Ω
1 kΩ
1.5 kΩ
2.5 kΩ
5 kΩ
10 kΩ
25 kΩ
50 kΩ
500 kΩ

Transformers, Inductors, and Coils

Center-tapped oscillator coil
30-mH choke coil
Air-core coils, 100 turns, A.S. No. 16 wire (2)
Air-core coil, 200 turns, A.S. No. 24 wire
Audio output transformer, 2.5-H primary

Capacitors (50 Vdcw unless specified)

50 μF
20 μF
1.5 μF
1 μF
0.1 μF, 200 Vdcw
0.01 μF
0.02 μF
24 pF
47 pF
20–150 pF (variable)

Semiconductors

Diodes: IN4004 (4)
Transistors: 2N2405 (2)
           2N1086
           2N3053 (2)
           GE-FET-1
Photocells: GE-X6 (3)
           GE-X19
           Photovoltaic solar cell
LED
LED readout: MAN-1
IC: 555 timers (2)
    LM3900 op amp
    RCA DR2000 Numitron

Equipment

Dual-channel oscilloscope
Digital vom
Ac-dc power supply

Vtvm
0–1000-mV meter
Zero-centered 50-mA meter
Zero-centered 100-mA, 500-Ω meter
Chart recorder
Decade resistance box
Signal generator

Specialized Components and Devices

Bimetal thermometer (0°F–250°F or
   −17.7°C–121.1°C)
Type J thermocouples (2)
GE-X15 thermistor
KA314 thermistor
JA41J1 thermistor
JA35J1 thermistor (2)
Rtd (100 Ω @ 0°C, platinum element)
Phototube: 917
Vacuum tube: 6C4
Optical coupler (Clairey CLM7H16A073)
Variable resistor with slider (250 Ω, 5 W)
Strain gages: foil, 120 Ω, GF = 2.1 (2)
Sling psychrometer (20°F to 120°F or −6.67°C
   to 48.8°C)

Tapered-tube flowmeter
Electrochemical cell: Plessey 560
Nixie tube with socket
Guardian 1335-2C-120D relay
U-tube manometer
Am receiver

Miscellaneous Components, Tools and Devices

Lamps with socket: 7 W, 60 W
Spst switch
660-W heat cone
Soldering gun
Styrofoam cups (2)
Cardboard box: $4 \times 4 \times 6$ inches ($10 \times 10 \times 15$ cm)
Cardboard cylinder: 1½ inches (D) $\times$ 6 inches or 3.8
cm (D) $\times$ 15 cm
Metal foil: 1×1 inch or 2.5×2.5 cm (2)
Steel core: ¾ (D) $\times$ 6 inches or 1.89×15 cm
Jar top
Adhesive
Balloon
Funnel
Plastic tubing
Plastic container

Contents

UNIT 1

Introduction to Instrumentation and Measurement Systems

UNIT 2

Thermal Input Transducers for Instrumentation Systems

UNIT 3

Photoelectric Transducers for Instrumentation Systems

UNIT 4

Miscellaneous Input Transducers for Instrumentation Systems

UNIT 5

Measurement of Physical Quantities

Introduction to Instrumentation and Measurement Systems

This unit provides a basic introduction to instrumentation and measurement. Activities 1-1 through 1-3 deal with measurement units, scientific notation, and metric measurement. An individual involved in instrumentation of any type must understand basic units. These activities make good review material for those already familiar with measurement. Activities 1-4 through 1-7 are used to familiarize the student with the measuring instruments which he or she will use in the laboratory. In most cases the student will be able to omit these activities, since most basic electricity and electronics courses deal with the use of specific types of vom's, electronic meters, oscilloscopes, and chart recorders.

Meter design using moving-coil movements is studied in Activities 1-8 through 1-13. These activities help the student to understand the operation of pointer-deflection meters such as vom's, vtvm's, and tvm's. They are also important for the student as a practical application of electrical circuit design. Several basic mathematical calculations are necessary to extend the range of a meter movement for measuring current, voltage, and resistance.

One of the most essential measurement circuits is the bridge circuit. This circuit is used to make comparative measurements where an unknown value is compared with a known value. The basic Wheatstone bridge and several other types of bridge circuits are studied in Activities 1-14 through 1-18. Another type of comparative measurement can be made with a potentiometric circuit used to measure voltage values. Activity 1-19 deals with potentiometric circuits. Frequency can also be measured through comparative techniques. Activity 1-20 investigates frequency measurement using Lissajous patterns on the oscilloscope.

Dew-point hygrometer indicator unit.

Courtesy General Eastern Instruments Corp.

Various types of numerical readout instruments are now used. These instruments provide a direct visual display of the value of the measured quantity; the user does not have to be able to interpret a meter scale. Activities 1-21 through 1-23 study numerical readout devices, such as Nixie tubes and LEDs, which are used with various instruments.

Introduction to Measurement

Introduction

Most measurement is based on the International System of Units (SI). The basic units of this system are the meter, kilogram, second, and ampere (MKSA). These are the units of length, mass, time, and electrical current. Other systems, such as the meter-kilogram-second (MKS) and centimeter-gram-second (CGS), recognize only three base units. These two systems, however, are closely associated with the International System.

There are several derived units that are used extensively for electrical and other related measurements. The electrical units that we now use are part of the International System of Units (SI) based on the meter-kilogram-second-ampere (MKSA) system. The International System of Derived Units is shown in Table 1-1A.

The coordination necessary to develop a standard system of electrical units is very complex. The International Advisory Committee on Electricity makes recommendations to the International Committee on Weights and Measures. Final authority is held by the General Conference on Weights and Measures, which meets periodically. The laboratory associated with the International System is the International Bureau of Weights and Measures, located near Paris, France. Several laboratories in different countries cooperate in this process of standardizing units of measurement. One such laboratory is the National Bureau of Standards in the United States.

Sometimes it is necessary to make conversions of measurement units so that very large or very small numbers may be avoided. For this reason, decimal multiples and submultiples of the basic units have been developed by using standard prefixes. These standard prefixes are shown in Table 1-1B. As examples, we may express 1000 volts as 1 kilovolt, or 0.001 ampere as 1 milliampere. A chart is shown in Fig. 1-1A that can be used for converting from one unit to another. To use this conversion chart, follow these simple steps:

1. Find the position of the unit as expressed in its original form.
2. Find the position of the unit to which you are converting.

Table 1-1A. International System of Derived Units

Measurement Quantity	SI Unit
Area	square meter
Volume	cubic meter
Frequency	hertz
Density	kilogram per cubic meter
Velocity	meter per second
Acceleration	meter per second per second
Force	newton
Pressure	pascal (newton per square meter)
Work (energy), quantity of heat....	joule
Power (mechanical, electrical)	watt
Electrical charge	coulomb
Permeability	henry per meter
Permittivity	farad per meter
Voltage, potential difference, electromotive force	volt
Electric flux density, displacement ...	coulomb per square meter
Electric field strength	volt per meter
Resistance	ohm
Capacitance	farad
Inductance	henry
Magnetic flux	weber
Magnetic flux density (magnetic induction)	tesla
Magnetic field strength (magnetic intensity).	ampere per meter
Magnetomotive force	ampere
Magnetic permeability	henry per meter
Luminous flux	lumen
Luminance	candela per square meter
Illumination	lux

3. Write the original number as a whole number and a derived unit or as a power of 10 and a basic unit.
4. Shift the decimal point the appropriate number of units in the direction of the term to which you are converting, and count the difference in decimal multiples from one unit to the other.

The use of this step-by-step procedure is illustrated in the following examples:

(a) Convert 100 picofarads to microfarads:
   1. 100 picofarads (pF).

   2. _____ microfarads (μF).
   3. 100 pF or 100×10^{-12}F.

Table 1-1B. Standard Prefixes

Prefix	Symbol	Factor by Which the Unit is Multiplied
exa	E	$1{,}000{,}000{,}000{,}000{,}000{,}000 = 10^{18}$
peta	P	$1{,}000{,}000{,}000{,}000{,}000 = 10^{15}$
tera	T	$1{,}000{,}000{,}000{,}000 = 10^{12}$
giga	G	$1{,}000{,}000{,}000 = 10^{9}$
mega	M	$1{,}000{,}000 = 10^{6}$
kilo	k	$1{,}000 = 10^{3}$
hecto	h	$100 = 10^{2}$
deka	da	$10 = 10^{1}$
deci	d	$0.1 = 10^{-1}$
centi	c	$0.01 = 10^{-2}$
milli	m	$0.001 = 10^{-3}$
micro	μ	$0.000001 = 10^{-6}$
nano	n	$0.000000001 = 10^{-9}$
pico	p	$0.000000000001 = 10^{-12}$
femto	f	$0.000000000000001 = 10^{-15}$
atto	a	$0.000000000000000001 = 10^{-18}$

4. 0.0001 μF (decimal shifted six units to the right).

(b) Convert 20,000 ohms to kilohms:
1. 20,000 ohms (Ω).

2. _____ kilohms (kΩ).
3. 20,000 Ω or $20{,}000 \times 10^{0}$ Ω.
4. 20 kΩ (decimal shifted three units to the left).

(c) Convert 10 milliamperes to microamperes:
1. 10 milliamperes (mA).

2. _____ microamperes (μA).
3. 10 mA or 10×10^{-3} A.
4. 10,000 μA (decimal shifted three units to the right).

Objective

In the study of instrumentation it is often necessary to convert numbers from one unit to another in order to solve problems. In this activity you will study unit conversion. You will then be able to successfully complete an exercise involving unit conversion.

Procedure

Convert each of the following quantities to the unit indicated. Use Table 1-1B and Fig. 1-1A to convert these units.
1. 0.0053 ampere to milliampere = _____ mA
2. 890 microamperes to amperes = _____ A
3. 3380 ohms to megohms = _____ MΩ
4. 0.65 megohm to ohms = _____ Ω
5. 1370 microvolts to millivolts = _____ mV
6. 16,000 watts to kilowatts = _____ kW

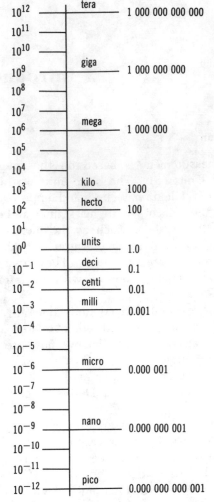

Fig. 1-1A. Conversion chart for large and small numbers.

7. 13,520,000 ohms to megohms = _____ MΩ
8. 25.24 volts to millivolts = _____ mV
9. 35,129 microamperes to amperes = _____ A
10. 83.21 milliamperes to microamperes = _____ μA
11. 2500 amperes to microamperes = _____ μA
12. 10,005 microamperes to milliamperes = _____ mA
13. 86,225 volts to kilovolts = _____ kV
14. 9122 ohms to kilohms = _____ kΩ
15. 218,000,000 ohms to megohms = _____ MΩ
16. 53.58 microvolts to volts = _____ V
17. 85,820 milliamperes to amperes = _____ A
18. 422 microamperes to amperes = _____ A
19. 58,250 watts to megawatts = _____ MW
20. 96.28 megohms to ohms = _____ Ω

Scientific Notation

Introduction

The technique of using powers of 10 can greatly simplify mathematical calculations. A number containing many zeros to the right or to the left of a decimal point can be dealt with much more easily when put in the form of powers of 10. For example, $0.0000027 \times 0.000028$ can be handled more easily when put in the form $(2.7 \times 10^{-6}) \times (2.8 \times 10^{-5})$. Notice the number of places that the decimal point is moved in each situation.

Table 1-2A. Powers of 10

Number	Power of 10
1,000,000	10^6
100,000	10^5
10,000	10^4
1000	10^3
100	10^2
10	10^1
1.0	10^0
0.1	10^{-1}
0.01	10^{-2}
0.001	10^{-3}
0.0001	10^{-4}
0.00001	10^{-5}
0.000001	10^{-6}

Table 1-2A lists some of the powers of 10. For a number consisting of a 1 followed by zeros, the power to which the 10 is raised is positive and equals the number of zeros following the 1. In decimal fractions (numbers less than 1), the power is negative and equals the number of places the decimal point is moved to the left.

Any number which is written as a product of a power of 10 and a number between 1 and 10 is said to be expressed in *scientific notation*. For example:

$$91,000,000 = 9.1 \times 10,000,000 \text{ or } 9.1 \times 10^7$$
$$800,000,000 = 8 \times 100,000,000 \text{ or } 8 \times 10^8$$
$$0.0000000007 = 7 \times 0.0000000001 \text{ or } 7 \times 10^{-10}$$

Scientific notation greatly simplifies the multiplication and division of large numbers or small decimals. For example:

$$1800 \times 0.000015 \times 300 \times 0.0048 = 1.8 \times 10^3 \times 1.5$$
$$\times 10^{-5} \times 3 \times 10^2$$
$$\times 4.8 \times 10^{-3}$$
$$= 1.8 \times 1.5 \times 3$$
$$\times 4.8$$
$$\times 10^{3-5+2-3}$$
$$= 38.88 \times 10^{-3}$$
$$= 0.03888$$

Divide 75,000 by 0.00005:

$$\frac{75,000}{0.00005} = \frac{7.5 \times 10^4}{5 \times 10^{-4}}$$
$$= \frac{7.5}{5} \times 10^{4-(-4)}$$
$$= 1.5 \times 10^8$$
$$= 150,000,000$$

Objective

Convert each of the following quantities to scientific notation. Place the correct responses in the blank spaces.

1. 0.0053 ampere $= 5.3 \times 10^{-3}$ A

2. 890 microamperes $= \underline{\hspace{2cm}} \mu$A

3. 3380 ohms $= \underline{\hspace{2cm}} \Omega$

4. 0.65 megohm $= \underline{\hspace{2cm}}$ MΩ

5. 1370 microvolts $= \underline{\hspace{2cm}} \mu$V

6. 16,000 watts $= \underline{\hspace{2cm}}$ W

7. 13,520,000 ohms $= \underline{\hspace{2cm}} \Omega$

8. 25.20 volts $= \underline{\hspace{2cm}}$ V

9. 35,200 microamperes $= \underline{\hspace{2cm}} \mu$A

10. 83.21 milliamperes $= \underline{\hspace{2cm}}$ mA

11. 2500 amperes = _____ A

12. 10.10 microamperes = _____ μA

13. 86,200 volts = _____ V

14. 9100 ohms = _____ Ω

15. 218,000,000 ohms = _____ Ω

16. 5350 microvolts = _____ μV

17. 85,300 milliamperes = _____ mA

18. 420 microamperes = _____ μA

19. 0.058 kilowatt = _____ kW

20. 0.962 megohm = _____ MΩ

Metric Measurements

Introduction

Most nations today use the metric system of measurement. In the United States the National Bureau of Standards began a study in August, 1968, to determine the feasibility and costs of converting industries and everyday activity to the metric system. Today this conversion is taking place. Metric measurement is becoming more common.

The units of the metric system are decimal measures based on the kilogram and the meter. Although this system is very simple, several countries have been slow to adopt it. The United States has been one of the reluctant countries due to the complexity of a complete changeover of measurement systems.

Objective

The importance of the metric system of measurement must be stressed in instrumentation. Conversion of units from the U.S. system to the metric system is easy to perform. In this activity you will become familiar with the metric system of measurement and complete a brief exercise dealing with metric conversions. Refer to the comparison of U.S. and metric units in Table 1-3A.

Procedure

1. List the basic units of the metric system.

   _____

   _____

   _____

2. Complete the following conversions:

   (a) 1 meter = _____ centimeters

   (b) 1 centimeter = _____ millimeters

   (c) 5000 grams = _____ kilograms

   (d) 2 liters = _____ milliliters

   (e) 1 meter = _____ inches

   (f) 1 centimeter = _____ inches

   (g) 1 mile = _____ kilometers

   (h) 1 gram = _____ ounces

   (i) 30 ounces = _____ grams

   (j) 1 kilogram = _____ pounds

   (k) 3 liters = _____ quarts

Table 1-3A. Comparison Chart for U.S. and Metric Units

U.S.		METRIC	METRIC		U.S.
1 inch	= 25.4	millimeters	1 millimeter	= 0.03937	inch
1 foot	= 0.3048	meter	1 meter	= 3.2808	feet
1 yard	= 0.9144	meter	1 meter	= 1.0936	yards
1 mile	= 1.609	kilometers	1 kilometer	= 0.6214	mile
1 square inch	= 6.4516	square centimeters	1 square centimeter	= 0.155	square inch
1 square foot	= 0.0929	square meter	1 square meter	= 10.7639	square feet
1 square yard	= 0.836	square meter	1 square meter	= 1.196	square yards
1 acre	= 0.4047	hectare	1 hectare	= 2.471	acres
1 cubic inch	= 16.387	cubic centimeters	1 cubic centimeter	= 0.061	cubic inch
1 cubic foot	= 0.028	cubic meter	1 cubic meter	= 35.3147	cubic feet
1 cubic yard	= 0.764	cubic meter	1 cubic meter	= 1.308	cubic yards
1 quart (liq)	= 0.946	liter	1 liter	= 1.0567	quarts (liq)
1 gallon	= 0.00378	cubic meter	1 cubic meter	= 264.172	gallons
1 ounce (avdp)	= 28.349	grams	1 gram	= 0.035	ounce (avdp)
1 pound (avdp)	= 0.4536	kilogram	1 kilogram	= 2.2046	pounds (avdp)
1 horsepower	= 0.7457	kilowatt	1 kilowatt	= 1.341	horsepower

(l)   1 gallon = _____ liters

(m) 10 cups = _____ liters

(n) 50 miles = _____ kilometers

(o) 2 cubic centimeters (cm³) = _____ cubic inches (in³)

(p) 4 cubic feet (ft³) = _____ cubic meters (m³)

Multifunction Meters

Introduction

Multifunction meters, such as the one shown in Fig. 1-4A, are used to measure more than one electrical quantity. Multifunction meters, such as the volt-ohm-milliammeter (vom), are used to measure voltage, resistance, and current. Thus the term *multifunction* is used to describe the meter. Such meters are also multirange meters since several ranges are provided for various levels of each electrical quantity to be measured.

The multifunction meter will be used to make several types of measurements in the laboratory activities that follow. You should obtain an operation manual for the meter and review it before using the meter.

Objective

In this activity you will review the operational characteristics of the multifunction meter that you will use in the laboratory. You will then answer several questions pertaining to this meter.

Equipment

Multifunction meter

Procedure

Complete the following statements or questions related to the operation of the multifunction meter you are using in the laboratory.

List the manufacturer of the meter: _____

List the model number of the meter: _____

1. The multifunction meter is scaled to measure up to _____ amperes of direct current.

2. Alternating current is read on the _____ colored scales.

3. The OHMS ADJUST control should be checked each time the resistance range is changed.   T   F

4. To measure a direct current greater than 2 amperes, the selector is placed in the _____ position.

5. For measuring current in a circuit, the meter should be connected in _____.

6. For measuring voltage, the meter should be connected in _____.

7. To measure 9 amperes of current, the selector must be placed in the _____ position.

8. To measure 10 microamperes of current, the selector must be placed in the _____ range.

9. To measure resistance, the red test lead must be placed in the jack marked _____ and the black test lead in the jack marked _____.

Courtesy Triplett Corp.

Fig. 1-4A. Multifunction meter (vom).

10. The most accurate resistance reading is found on the _____ side of the meter scale.
11. Polarity is important when measuring ac voltage.　T　F
12. Polarity is not important when measuring resistance.　T　F
13. Up to _____ volts dc can be measured with the multifunction meter.
14. For measuring a resistor valued at 10 ohms, the _____ range would be used.
15. The top scale of the multifunction meter is used to measure ac voltage.　T　F
16. Polarities must be observed when measuring dc.　T　F
17. It is considered good practice to measure the resistance of a circuit with voltage applied to the circuit.　T　F
18. When measuring an unknown value of dc voltage, one should start at the highest scale and work down to the correct scale.　T　F
19. The meter used in the lab is a multifunction and also a multirange meter.　T　F
20. The multifunction meter has an overload circuit designed to protect the meter from damage.　T　F

Electronic Multifunction Meters

Introduction

Electronic multifunction meters are used to make measurements in circuits where the "loading" effect of a self-powered meter may cause erroneous readings. The input impedance of an electronic multifunction meter, such as a digital voltmeter (dvm), transistorized voltmeter (tvm), field-effect transistor meter (fetm), or vacuum-tube voltmeter (vtvm), is ordinarily much higher than that of a vom or similar self-contained meter. The disadvantage of some types of these meters is that an ac voltage source is required; therefore, they are not as portable as the vom. Some types of electronic multifunction meters are shown in Fig. 1-5A.

Objective

In this activity you will review the operational characteristics of the electronic multifunction meter that you will use in the lab. You should obtain a manual of operation for the meter and review it before using the meter.

Equipment

Multifunction meter and manual

Procedure

Complete each of the following statements that pertain to the operation of the electronic multifunction meter in your lab. If the statement is true, false, or not applicable, circle the T, F, or NA.

List the manufacturer of the meter: _____

List the model number of the meter: _____
List the specific type of meter you are using, such as

vtvm, tvm, or tvom. _____

1. Resistance is read on the top scale of the meter. T F NA

2. The red colored scales are used to measure ac voltage. T F NA

3. Current, either ac or dc, can be measured with the meter. T F

4. There are several quantities which may be measured with the meter.

   They are: _____

5. The probe of the vtvm is considered negative when measuring dc. T F

6. Peak-to-peak voltages may be read directly on the meter. T F

7. When resistance is to be measured, all power in the external circuit must be turned off. T F

8. The meter can be "zero centered." T F

9. The meter can measure both positive and negative dc voltages. T F

10. When resistance is measured, the pointer deflects from left to right. T F NA

11. When any value is to be measured, one should start with the highest range and work down to the proper scale. T F

12. The ohms adjust control must be readjusted each time a new range is selected for measuring resistance. T F

13. When dc voltage is measured, polarity must be observed. T F

14. Polarity is not important when ac voltage is measured. T F

15. The meter has an overload circuit designed to protect it. T F

(a) Transistorized volt-ohm-milliammeter (tvom).

(b) Vacuum-tube voltmeter (vtvm).

(c) Transistorized voltmeter (tvm).

Courtesy Triplett Corp.

Fig. 1-5A. Electronic multifunction meters.

Oscilloscope Measurements

Introduction

The oscilloscope is an important type of instrument. It is possible to monitor voltages visually by using an oscilloscope. The basic operational part of the oscilloscope is the cathode-ray tube (crt). You should become familiar with the operation of the cathode-ray tube in order to better understand the oscilloscope.

Fig. 1-6A shows the construction details of a crt, while Fig. 1-6B illustrates the operation of the electron-gun assembly in the neck of the crt. A beam of electrons is produced by the cathode of the tube when it is heated by the application of a filament voltage. The electrons are attracted to the positive potential of anode No. 1. The quantity of electrons passing on to anode No. 1 is determined by the amount of nega-

tive bias voltage applied to the control grid. Anode No. 2 is operated at a higher positive potential to further accelerate the electron beam toward the screen of the crt. The difference in potential between anode No. 1 and anode No. 2 determines the point of convergence of the beam on the crt screen. When the electron beam strikes the phosphorescent screen of the crt, light is emitted.

To control the horizontal and vertical movement of the electron beam, deflection plates are used. If no potential is applied to either plate, the electron beam would merely appear as a dot in the center of the crt screen. Electrostatic deflection occurs when a potential is placed on the horizontal and vertical deflection plates. Horizontal deflection usually results from a sawtooth waveform produced by a sweep oscillator,

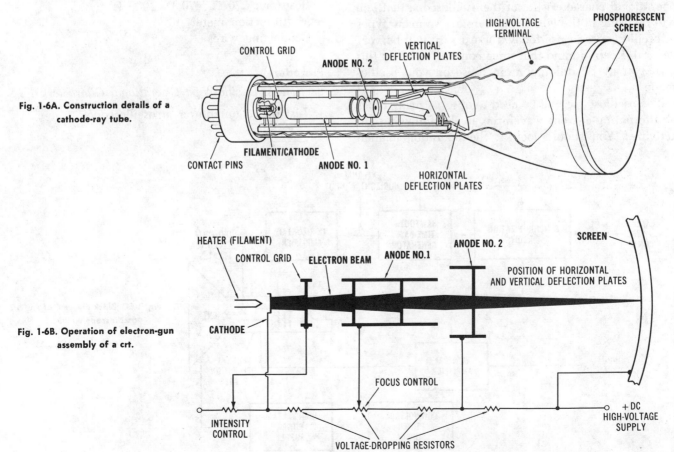

Fig. 1-6A. Construction details of a cathode-ray tube.

Fig. 1-6B. Operation of electron-gun assembly of a crt.

which is part of the oscilloscope circuit. This voltage "sweeps" the electron beam back and forth across the crt screen. The sweep setting of a crt is adjusted to coincide with the frequency range of the voltage to be measured. Vertical deflection takes place in accordance with the magnitude of the applied voltage. The voltage to be measured is applied to the vertical deflection circuit of the oscilloscope (see Fig. 1-6C).

Oscilloscopes are available in several different types. General-purpose oscilloscopes are used to display simple types of waveforms, and for general electronic servicing. Triggered-sweep oscilloscopes are used where it is desired to apply external voltage to the horizontal circuits to produce horizontal sweep. Other oscilloscopes, classified as laboratory types, have very high sensitivity and good frequency response over a wide range. The crt display indicators can be used to measure ac and dc voltages, frequency and phase relationships, distortion in amplifiers, and various timing and numerical-control applications. Memory and storage-type instruments are also available for more sophisticated measurements.

Objective

In Section A of this activity you will review the operational characteristics of the oscilloscope that you will be using in the lab. Since there are so many types of oscilloscopes in use today, the exercise will be very general. Keep in mind that some controls listed on this activity may be different from those on your oscilloscope.

The oscilloscope will be used to observe many types of alternating-current waveforms in the following lab activities. You should look at the operation manual for the scope you are using, if it is available, before completing the activity.

In Section B of this activity you will use the oscilloscope to measure the amplitude of an ac sine-wave voltage. The voltage amplitude of an ac sine wave can be measured in three different ways: (1) peak-to-peak (p-p), (2) peak (pk), and (3) root-mean-square (rms), or effective, values. The *peak* (pk) value is simply the maximum value, measured from zero, of either the positive or negative half. The *peak-to-peak* value is the voltage difference between the peak positive and peak negative values. The *rms* (or *effective*) voltage of a waveform is the value of that dc voltage which will do the same amount of work as the original waveform. For a sine wave the effective value is 0.707 of the peak value. The power-line voltage in the home, for example, is stated as an rms value. Most ac meters accurately measure only sine-wave voltages, but the oscilloscope can measure all voltages regardless of waveform type. This is a primary advantage of the oscilloscope.

Equipment

Oscilloscope
Ac power source
Resistors: 270 Ω, 470 Ω, 1000 Ω
Multifunction meter
Connecting wires

Procedure

Section A: Oscilloscope Operational Characteristics

1. List the manufacturer of oscilloscope:

_____

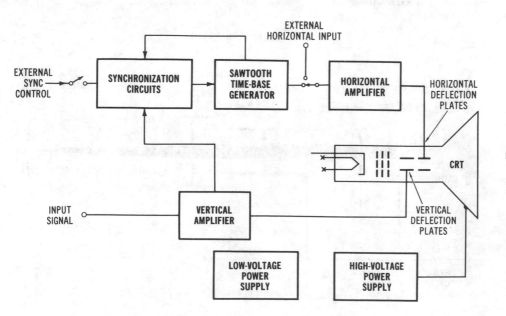

Fig. 1-6C. Block diagram of an oscilloscope circuit.

2. List the model number of oscilloscope:

_____

3. In the following, match the oscilloscope control with the proper function. If the control on your scope has a different name, place it in parentheses beside the control name listed.

_____ 1. INTENSITY

_____ 2. FOCUS

_____ 3. VERTICAL POSITION

_____ 4. HORIZONTAL POSITION

_____ 5. VERTICAL GAIN

_____ 6. HORIZONTAL GAIN

_____ 7. VERTICAL ATTENUATION

_____ 8. SWEEP SELECT

_____ 9. VERNIER

_____ 10. SYNCHRONIZATION SELECT

_____ 11. HORIZONTAL INPUT

_____ 12. VERTICAL INPUT

(a) Selects horizontal sweep frequency
(b) Provides up-and-down adjustment of beam
(c) Determines amplitude of horizontal deflection
(d) Controls size of electron beam
(e) Provides left-and-right adjustment of beam
(f) Controls brightness of beam
(g) Eliminates vertical shift of display
(h) External voltages applied to this control
(i) Determines frequency of the sawtooth sweep
(j) Determines amplitude of vertical deflection
(k) Setting for calibration voltage
(l) Controls external signals to vertical amplifier
(m) Determines amount of voltage used to synchronize sweep oscillator
(n) Selects type of synchronization desired
(o) Balances dc output
(p) Sweeps at line frequency

Section B: Measuring Sine-Wave Amplitude

1. Obtain a variable ac power supply and adjust the voltage to 10 volts ac. Calculate the peak and peak-to-peak values of this voltage.

Peak value = _____ volts ac

Peak-to-peak value = _____ volts ac

2. Construct the circuit shown in Fig. 1-6D and connect it to the 10-volt ac source. Measure the voltage across the following points with the meter. Calculate the peak (pk) and peak-to-peak (p-p) values.
(a) Points A to B:

_____ volts rms; _____ volts pk;

_____ volts p-p
(b) Points B to C:

_____ volts rms; _____ volts pk;

_____ volts p-p
(c) Points C to D:

_____ volts rms; _____ volts pk;

_____ volts p-p
3. Prepare the oscilloscope for operation by adjusting the appropriate controls. Connect the ground lead to point D and the vertical-input lead to point A. Properly center the waveform and adjust the horizontal frequency and sync controls to produce two sine waves. Adjust the vertical controls so that

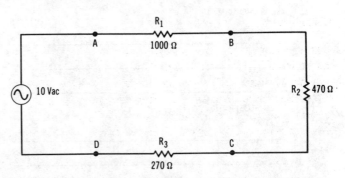

Fig. 1-6D. Test circuit for ac voltage measurements.

each square or horizontal line of the graticule on the scope screen equals a definite peak-to-peak voltage. As previously determined mathematically, the peak-to-peak voltage of the 10-volt source is approximately 28.2 volts. Calibrate the scope graticule so the sine wave has a specific number of squares or lines of amplitude. Once the scope has been calibrated, do not adjust the vertical controls again. (NOTE: Some scopes have internal calibration and the preceding is not required.)

4. With the scope calibrated, measure the peak-to-peak voltage across the following points and record the values:

(a) Points A to B = _____ volts p-p

(b) Points B to C = _____ volts p-p

(c) Points C to D = _____ volts p-p

5. This concludes the activity.

Analysis

1. Define the following terms that are related to oscilloscope use with alternating-current signals applied.

   (a) Cycle:

   _____

   _____

   _____

   (b) Frequency:

   _____

   _____

   (c) Hertz:

   _____

   _____

   _____

   (d) Rms voltage:

   _____

   _____

   _____

   (e) Peak-to-peak voltage:

   _____

   _____

   _____

   (f) Kilohertz:

   _____

   _____

   (g) Megahertz:

   _____

   _____

   _____

   (h) Amplitude:

   _____

   _____

   _____

2. In order to measure peak-to-peak voltage, the scope must first be calibrated, then it can be used to make any peak voltage measurements. How does changing the VERTICAL ATTENUATION or VOLTS/DIVISION control affect the calibration?

   _____

   _____

3. Discuss the basic operation of the oscilloscope.

   _____

   _____

   _____

   _____

   _____

   _____

4. How can an oscilloscope be used to measure frequency?

   _____

   _____

5. Explain the procedure for calibrating an oscilloscope to measure peak-to-peak ac voltage.

   _____

   _____

   _____

Chart Recording Instruments

Introduction

Chart recorders are used when a permanent record of a measured quantity is needed. Instruments of this type are employed to provide a permanent record of some quantity that is measured over a specific period. A typical chart-recording instrument is shown in Fig. 1-7A. Types of chart recorders include pen-and-ink recorders and inkless recorders.

Pen-and-ink recorders use a pen which touches a paper chart and leaves a permanent record of the measured quantity on the chart. The charts utilized can either be roll charts, which revolve on rollers under the pen mechanism, or circular charts, which revolve on an axis under the pen. Chart recorders may use more than one pen to record several different quantities simultaneously.

The pen of a chart recorder is a capillary-tube device. The pen must be connected to a constant source of ink. The pen is moved by the torque exerted by the meter movement. The chart used for recording the measured quantity usually contains lines that correspond to the radius of the pen movement. Increments on the chart are marked according to time intervals. The chart must be moved under the pen at a constant speed. Either a spring-drive mechanism, synchronous ac motor, or a dc servomotor could be used to drive the chart. Recorders are also available that use a single pen to make permanent records of measured quantities on a single chart. In this case either coded lines or different-colored ink could be used to record the quantities.

Inkless recorders may use a voltage applied to the pen point to produce an impression on a sensitive paper chart. In another process the pen is heated to cause a trace to be melted along the chart paper. The obvious advantage of inkless recorders is that ink is not required.

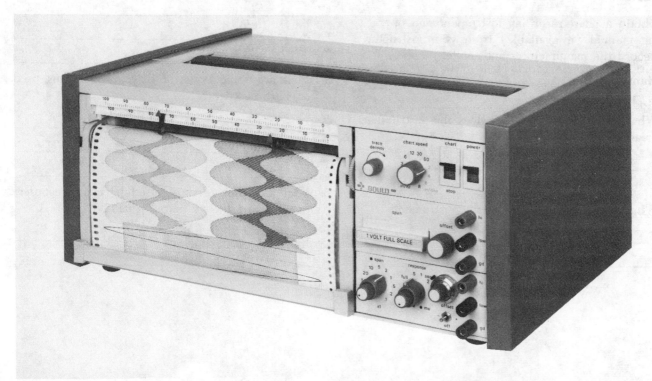

Courtesy Gould, Inc.

Fig. 1-7A. Chart-recording indicator.

Chart-recording indicators for measuring almost any electrical or physical quantity are commerically available. For many applications the recording system may be located some distance from the quantity being measured. For accurate process control or system monitoring a central instrumentation system might be used. Power plants, for instance, ordinarily use chart recorders at a centralized location to monitor the various electrical and physical quantities involved in the power-plant operation.

Objective

Chart recorders are used to monitor many types of electrical and physical quantities over a period. The advantage of a chart recorder is that a visible record of the monitored quantity may be kept for future use. There are many different types of chart recorders in use today.

In this laboratory activity you will be able to study the construction and operation of a chart recorder. The recorder should be connected to some operational system so that you can observe its operation.

Equipment

Chart recorder (any type available in the lab)

Procedure

1. Obtain a chart recording instrument and operating manual (if available) from your instructor. Record the manufacturer and model number.

   Manufacturer: _____

   Model number: _____
2. What quantity or quantities does this instrument measure?

   _____

   _____
3. What is the range or scale of this instrument?

   _____

   _____

4. What method is used by this instrument to make a permanent record of the quantity measured?

   _____
5. What type of drive mechanism is used on this instrument?

   _____
6. What type of chart is used by this instrument?

   _____
7. Describe any special features this instrument has.

   _____

   _____
8. Prepare the chart recorder for operation. Connect the recorder to some circuit or system in which you can observe its operation over its entire range of measurement.
9. Have your instructor check the instrument for operation.

   Instructor's approval: _____
10. This concludes the laboratory activity.

Analysis

1. What are some specific applications where chart recorders might be used?

   _____

   _____

   _____
2. What are some types of drive mechanisms used for chart recorders?

   _____

   _____

   _____
3. What are some methods used by chart recorders for recording the measured quantity?

   _____

   _____

Instrument Meter Movements

Introduction

Instruments that rely on the motion of a hand or pointer assembly are referred to as *hand-deflection* meters. The volt-ohm-milliammeter (vom) is one type of hand-deflection meter. The vom is a multifunction instrument that is used for measuring electrical quantities. Single-function hand-deflection indicators are also commonly used to measure electrical or physical quantities. The basic part of an electrical hand-deflec-

tion meter is called a *meter movement*. Physical quantities such as air flow or fluid pressure can also be monitored by hand-deflection meters. The movement of the hand or pointer relative to a calibrated scale indicates the electrical or physical quantity.

Many meters employ the *D'Arsonval* type of meter movement. Construction details and the basic operational principle of this type of movement are shown in Fig. 1-8A. The hand or pointer of the movement remains stationary on the left portion of the calibrated

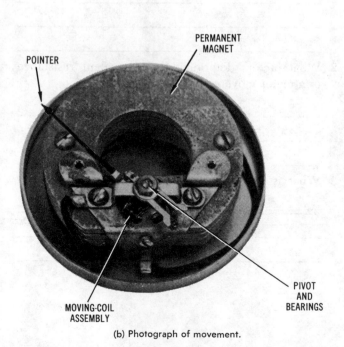

(a) Construction details.

(b) Photograph of movement.

(c) Operating principle.

Fig. 1-8A. D'Arsonval type of hand-deflection meter movement.

scale until current flows through the electromagnetic coil that is centrally located within a permanent magnetic field. When current flows through the coil, a reaction between the electromagnetic field of the coil and the stationary permanent magnetic field is developed. This reaction causes the hand to deflect toward the right portion of the scale. This basic moving-coil meter movement operates on the same principle as an electric motor. It can be used for either single-function electrical instruments, which measure only one quantity, or for multifunction meters such as a vom or vtvm (vacuum-tube voltmeter). The basic D'Arsonval meter movement can be modified so that it will measure voltage, current, or resistance.

Objective

In this laboratory activity you will work with a basic meter movement which is used in many types of instruments. You will employ an experimental method to determine the equivalent resistance of the movement.

Equipment

Meter movement: 0 to 100 μA, 500 Ω or equivalent
Multifunction meter
Potentiometer: 1000 Ω
Variable dc power supply
Resistor: 10 kΩ, 2 W
Connecting wires

Procedure

1. Obtain a 0–100-μA, zero-centered meter movement.
2. Use the circuit shown in Fig. 1-8B to determine the resistance of the meter movement. This value cannot be measured directly with an ohmmeter.
3. Connect the meter movement in series with a 10,000-Ω resistor and dc power source as shown, with R_{SH} out of the circuit. Slowly increase the dc voltage until the meter reads full scale. Now connect the potentiometer (R_{SH}) in parallel with the meter movement. Do not disturb the dc power supply setting.
4. Adjust the potentiometer until the meter movement reads half scale. Turn off the power, remove the potentiometer, and without disturbing its adjust-

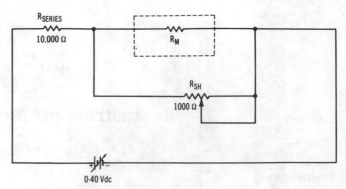

Fig. 1-8B. Circuit for determining the resistance of a meter movement.

ment, measure its resistance. This resistance is equal to the meter movement resistance (R_M).

$$R_M = \underline{\hspace{1cm}} \ \Omega$$

5. Using the meter movement resistance (R_M) and the full-scale current (I_{FS}), write the meter movement specification:

Movement specification $= I_{FS} \times R_M = \underline{\hspace{1cm}}$ mV

Analysis

1. Describe the construction of a meter movement such as the one used in this activity.

_____

_____

_____

_____

_____

_____

2. What factors determine the equivalent resistance of a meter movement?

_____

_____

3. What is meant by meter movement specification?

_____

_____

_____

Meter Sensitivity

Introduction

An important aspect of meter operation is called *sensitivity*. The sensitivity of a meter increases as the current for full-scale deflection (I_M) decreases. Since a voltmeter is connected in parallel with the circuit under test, part of the circuit current flows through the meter. The current that flows through the meter should be small compared to the circuit current. The more sensitive the meter, the less current it draws from the circuit. Sensitivity is expressed in ohms per volt, and is equal to $1/I_M$, where I_M is the current required for full-scale deflection of the meter. A 50-μA movement would have an ohms-per-volt rating of 1/0.000005, or 20,000 ohms per volt. This means that the meter would require 20,000 ohms for a 1-volt range. Sensitivities of meter movements range from as low as 100 ohms per volt to as high as 200,000 ohms per volt. Low-sensitivity meters should not be used for making measurements in circuits that have small currents. Electronic meters have sensitivities in the megohm range. For example, a tvm might have a sensitivity of 10 megohms per volt. The larger values of sensitivity are said to have less "loading" effect on a circuit being tested. "Loading" refers to erroneous meter readings caused when the meter draws a large current from the circuit being measured. You can see that the larger the ohms-per-volt rating of a meter, the smaller is the current through the meter to produce full-scale deflection.

Objective

In this laboratory activity you will study the sensitivity of meter movements and the relationship of sensitivity and circuit loading effect.

Equipment

Voltmeter (any type with 10,000 ohms-per-volt sensitivity or lower)
Electronic multifunction meter (with 1 megohm or higher sensitivity)
Variable dc power supply

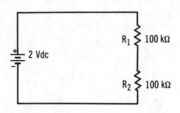

Fig. 1-9A. Circuit illustrating meter loading effect.

Resistors: 100 kΩ (2)
Connecting wires

Procedure

1. Construct the circuit shown in Fig. 1-9A.
2. Calculate the voltage drops across R_1 and R_2.

   $V_{R1} =$ _____ Vdc; $V_{R2} =$ _____ Vdc
3. Measure the voltage drops across R_1 and R_2 using a voltmeter which has a sensitivity of 10,000 ohms per volt or less. Record these readings.

   $V_{R1} =$ _____ Vdc; $V_{R2} =$ _____ Vdc
4. Now measure the voltage drops across R_1 and R_2 using an electronic voltmeter which has a sensitivity of 1 megohm per volt or higher. Record these voltages.

   $V_{R1} =$ _____ Vdc; $V_{R2} =$ _____ Vdc
5. This concludes the activity.

Analysis

1. Which meter used in the activity produces a more accurate measurement of voltage? Why?

   _____

   _____

   _____

2. Calculate the percent difference between values of V_{R1} from Steps 2 and 3.

   % Difference =
   $$\frac{\text{Calculated Value} - \text{Measured Value}}{\text{Calculated Value}} \times 100$$

   = _____ %

3. Calculate the percent difference between values of V_{R1} from Steps 2 and 4.

   % Difference = _____ %

4. What is meant by meter *sensitivity?*

   _____

   _____

   _____

5. What is meant by meter *loading?*

   _____

   _____

   _____

Direct-Current Measurement

Introduction

The moving-coil meter can be used to measure various values of direct current by employing shunt resistors in parallel with the movement. A schematic of the basic movement is illustrated in Fig. 1-10A. The coil of the meter movement is not designed to handle high currents. The typical current rating (I_M) of a meter movement ranges from 10 μA to 10 mA. The current rating indicates the current required to cause the pointer to deflect to the full-scale (extreme right) position. When the current value is lowered, more turns of wire must be used for the coil in order to develop sufficient electromagnetic field strength.

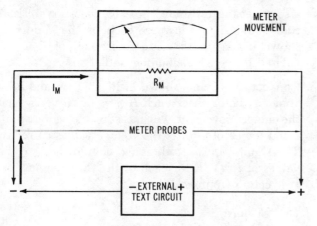

Fig. 1-10A. Basic meter movement.

Therefore, as the current rating (I_M) decreases, the resistance (R_M) of the movement increases. We must know the values of I_M and R_M in order to design a meter circuit that will measure currents higher than the value of I_M.

If a resistor is placed in parallel with the basic meter movement, as shown in Fig. 1-10B, part of the current from the external circuit is shunted through the resistor. Therefore this resistor is called a *shunt* resistor (R_{SH}). When properly calculated, the value of shunt resistance will establish meter circuit conditions so that other ranges of current (above I_M) can be measured by the same movement. The value of shunt resistance can be expressed as:

$$R_{SH} = \frac{I_M \times R_M}{I_{SH}}$$

where

R_{SH} is the shunt resistance in ohms,
I_M is the full-scale current rating of the meter movement in amperes,
R_M is the resistance of the meter movement in ohms,
I_{SH} is the desired value of current through the shunt resistance in amperes.

For example, if we have a 1.0-mA, 50-Ω meter movement and wish to design a meter to measure 10 mA, we can determine the shunt resistance as follows:

$$\begin{aligned}
R_{SH} &= \frac{I_M \times R_M}{I_{SH}} \\
&= \frac{0.001 \text{ A} \times 50 \text{ }\Omega}{0.009 \text{ A}} \\
&= 5.55 \text{ }\Omega
\end{aligned}$$

The value of I_{SH} was set at 9 mA (0.009 A) since that amount of current had to be shunted so that at the maximum current of the extended range, I_M would still equal 1.0 mA (9 mA + 1.0 mA = 10 mA). A multirange ammeter can be designed by using several values of shunt resistance and a switching arrangement as shown in Fig. 1-10C. Multirange meters have

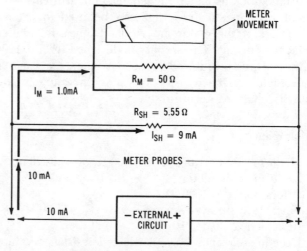

Fig. 1-10B. Meter movement with shunt resistor.

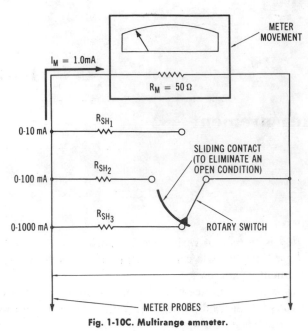

Fig. 1-10C. Multirange ammeter.

only one scale; however, either a directly read scale or a scale with a multiplying factor is provided for ease of reading each range.

Objective

The basic meter movement is a current-sensitive device. A 0–5-mA meter movement, for instance, requires 5 mA of current to register full-scale deflection. A current of 2.5 mA would cause one-half scale deflection. If values of current larger than 5 mA were to be measured, this basic movement could not be used. However, the range of the meter can be extended to measure higher current values by adding a *shunt* resistor of a specified value in parallel with the movement.

In this laboratory activity you will determine the resistance of the meter movement and then calculate shunt resistance values that can be used to extend the range of the meter movement to measure higher values of current. The activity is designed for a 0–100-μA, 500-Ω, zero-centered meter movement. Any meter movement can be used, however, with values in the activity changed accordingly.

Equipment

Meter movement: 0–100 μA, 500 Ω, or equivalent
Multifunction meter
Potentiometer: 1000 Ω
Variable dc power source
Resistors: 1200 Ω, 10 W
Decade resistance box or selected resistance values
Connecting wires

Procedure

1. Using the 100-μA meter movement and its resistance (R_M), calculate the values of shunt resistance (R_{SH}) needed to extend the range of the movement to measure:

(a) 1.0 mA; $R_{SH} = $ _____ Ω

_____

_____

_____

(b) 10 mA; $R_{SH} = $ _____ Ω

_____

_____

_____

(c) 100 mA; $R_{SH} = $ _____ Ω

_____

_____

_____

2. Use a decade resistance box, or necessary resistor combinations, to construct each of the shunts as shown in Fig. 1-10D. Use the values of R_{SH} calculated in Step 1, beginning with 1 mA.

3. Connect the circuits shown in Fig. 1-10D, and have your instructor check each R_{SH} value to verify that the meter movement will measure the values indicated. Be careful to set the voltage values correctly. The shunts you have designed should indicate the values of $I = V/R$ for each of the three circuits. Instructor's approval:

(a) 1.0 mA: _____

(b) 10 mA: _____

(c) 100 mA: _____

Analysis

1. If a 200-Ω, 1.0-mA meter movement is employed, what are the values of shunt resistance required to extend its range to measure the following currents? (Show your calculations.)

(a) 5 mA; $R_{SH} = $ _____ Ω

_____

_____

_____

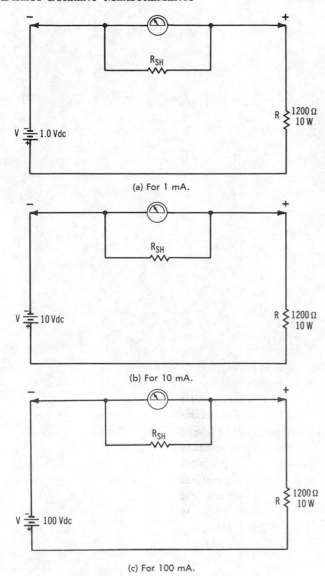

(a) For 1 mA.

(b) For 10 mA.

(c) For 100 mA.

Fig. 1-10D. Meter shunt test circuits.

(b) 10 mA; $R_{SH} = $ _____ Ω

_____

_____

(c) 50 mA; $R_{SH} = $ _____ Ω

_____

_____

(d) 100 mA; $R_{SH} = $ _____ Ω

_____

_____

(e) 0.5 A; $R_{SH} = $ _____ Ω

_____

_____

(f) 1.0 A; $R_{SH} = $ _____ Ω

2. What factors determine the R_M value of a meter movement?

_____

_____

_____

DC Voltage Measurement

Introduction

Meter movements can also be used to measure dc voltage. Although the basic movement actually responds to current through its electromagnetic coil, predetermined voltages can also produce accurate readings on a calibrated meter scale. To measure voltage a resistor is placed in series with the meter movement. This resistor, shown in Fig. 1-11A, is called a *multiplier*. The purpose of the multiplier is to adjust the value of current in the meter circuit so that at a predetermined voltage, the meter current will cause full-scale deflection on the meter scale. This is true since

$$V_{FS} = I_M \times R_M$$

where

V_{FS} is the voltage required for full-scale deflection in volts,

I_M is the current required for full-scale deflection in amperes,

R_M is the resistance of the meter movement in ohms.

Assume that we wish to measure a range of dc voltage from 0 to 10 V using a 1.0-mA, 50-Ω meter movement. With 1.0 mA through the 50-Ω meter movement

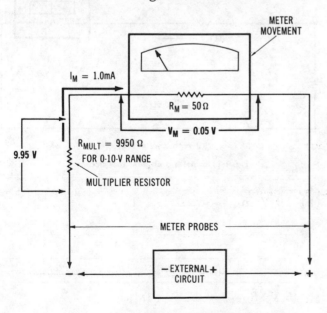

METER MOVEMENT

$I_M = 1.0\text{mA}$

$R_M = 50\,\Omega$

$V_M = 0.05\,\text{V}$

9.95 V

$R_{MULT} = 9950\,\Omega$
FOR 0-10-V RANGE

MULTIPLIER RESISTOR

METER PROBES

− EXTERNAL +
CIRCUIT

Fig. 1-11A. Measuring dc voltage.

at full-scale deflection, there must be 0.05 V applied across the meter terminals, as shown in Fig. 1-11A. Thus, in order to measure 10 volts at full-scale deflection, we must add sufficient multiplier resistance in series with the meter to drop 9.95 volts with a 1.0-mA current in the circuit, or:

$$R_{MULT} = \frac{V_{MULT}}{I_T}$$
$$= \frac{9.95\text{ V}}{0.001\text{ A}}$$
$$= 9950\,\Omega$$

Another way to determine the value of the multiplier resistance is to say that the total resistance in the combined circuit with 10 V applied is:

$$R_T = \frac{V_T}{I_T}$$
$$= \frac{10\text{ V}}{0.001\text{ A}}$$
$$= 10,000\,\Omega$$

Since the total resistance of the circuit is also $R_T = R_M + R_{MULT}$, then:

$$R_{MULT} = R_T - R_M$$
$$= 10,000\,\Omega - 50\,\Omega$$
$$= 9950\,\Omega$$

This figure is the same as calculated value.

A switching arrangement can be used to provide several voltage ranges using the same meter movement.

Objective

In this activity you will use a basic meter movement to measure dc voltage. In the previous activity the meter movement was used to measure current. In a similar manner the basic meter movement can be used to measure voltage values. When a voltage is applied to the meter movement, a certain amount of current flows through the movement. The voltmeter movement can also be extended to measure higher values; in this case a *multiplier* resistor is placed in series with the movement.

Equipment

Meter movement: 0–100 μA, zero-centered, 500 Ω,
  or equivalent
Variable dc power source
Decade resistance box or assorted resistors
Multifunction meter
Connecting wires

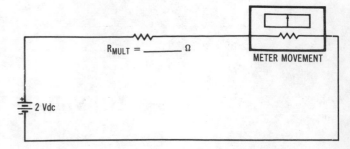

(a) With 2-V source.

Procedure

1. The circuit shown in Fig. 1-11B illustrates how a
meter movement is used to measure voltages.

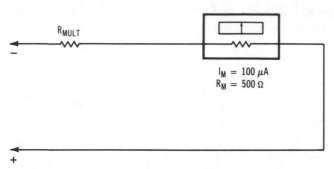

$$I_M = 100 \ \mu A$$
$$R_M = 500 \ \Omega$$

Fig. 1-11B. Basic meter movement for measuring voltages.

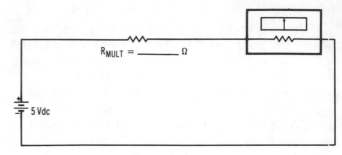

(b) With 5-V source.

2. Using the 0–100-μA, 500-Ω meter movement, calculate the value of multiplier resistance needed to
extend the meter range to measure the following
voltages:

(a) 2 V; $R_{\text{MULT}} =$ _____ Ω

(b) 5 V; $R_{\text{MULT}} =$ _____ Ω

(c) 10 V; $R_{\text{MULT}} =$ _____ Ω

(d) 25 V; $R_{\text{MULT}} =$ _____ Ω

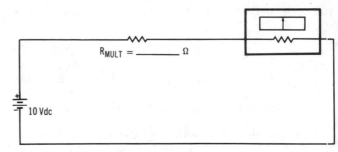

(c) With 10-V source.

3. Have your instructor check these values as you
verify them by applying the indicated voltages to
the meter circuits of Fig. 1-11C with your calculated values of R_{MULT}. *Be careful.*
Instructor's approval:

(a) 2 V _____

(b) 5 V _____

(c) 10 V _____

(d) 25 V _____

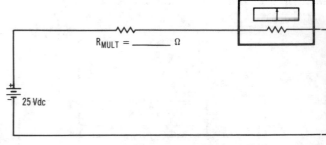

(d) With 25-V source.

Fig. 1-11C. Meter multiplier test circuits.

Analysis

Calculate the values of multiplier resistance needed
to extend the range of a 1.0-mA, 100-Ω meter movement to measure the following voltages:

(a) 1.0 V; $R_{\text{MULT}} =$ _____ Ω

_____

_____

_____

(b) 5 V; $R_{\text{MULT}} =$ _____ Ω

_____

_____

_____

(c) 15 V; $R_{\text{MULT}} =$ _____ Ω

_____

_____

_____

(d) 150 V; $R_{\text{MULT}} =$ _____ Ω

_____

_____

(e) 200 V; $R_{\text{MULT}} =$ _____ Ω

_____

_____

_____

(f) 500 V; $R_{\text{MULT}} =$ _____ Ω

_____

_____

_____

Resistance Measurement

Introduction

Resistance can also be measured by using the basic D'Arsonval meter movement. One type of ohmmeter circuit is shown in Fig. 1-12A. This circuit is called a *series* ohmmeter since the resistance to be measured is connected in series with the meter circuit. Note that a 1.5-V cell is used to supply current to the meter circuit. The scale of the meter is calibrated so that the deflection of the pointer will indicate a specific value of resistance. When the meter probes are shorted together (zero external resistance), the meter pointer is adjusted to read full-scale. The full-scale mark on the "ohms" scale, shown in Fig. 1-12B, indicates zero resistance. As higher values of resistance are measured between the meter probes, less current will flow through the ohmmeter circuit. Thus the meter pointer will deflect less for higher resistance values. The OHMS ADJUST control is a variable resistance in the ohmmeter circuit that is used to adjust the meter hand to zero on the scale when the probes are shorted together. The purpose of the OHMS ADJUST control is to compensate for any variation in the voltage of the battery used to supply current to the circuit. The other side of the scale indicates infinite resistance (open circuit).

In the sample ohmmeter circuit shown in Fig. 1-12A, a 1.0-mA, 50-Ω meter movement is used. For our dis-

cussion we will assume a value of 500 Ω for the OHMS ADJUST control and a value of 1200 Ω for the current-limiting resistor, R_{MULT}. With the OHMS ADJUST control set to a value of 250 Ω the total resistance of the meter circuit will be $R_M + R_{MULT} + R_{OHMS}$, or 50 + 1200 + 250 = 1500 Ω. Then, when we short the meter probes together, the total current in the circuit will be:

$$I_T = \frac{V_T}{R_T}$$
$$= \frac{1.5 \text{ V}}{1500 \text{ }\Omega}$$
$$= 0.001 \text{ ampere } (1.0 \text{ mA})$$

which is the current value required for full-scale deflection of the meter. The "ohms" scale is then calibrated to read zero ohms at the point of full-scale deflection.

The external resistance that would cause half-scale deflection of the meter pointer would be equal to the total resistance of the ohmmeter circuit that limits the current to full-scale deflection (1500 ohms). With 1500 ohms external resistance connected between the meter probes, the total circuit resistance will be 3000 ohms. The meter pointer will then deflect to only one-half of full scale since, with twice as much resistance in the circuit, there will be only half as much current. Therefore we can mark the center of the ohms scale for 1500 ohms as shown in Fig. 1-12B.

If we connect 3000 Ω external resistance between the meter probes, the total circuit resistance will be 4500 Ω, or three times the resistance of the meter circuit. The total current in the circuit then will be only

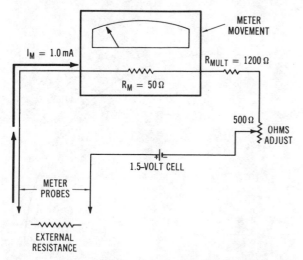

Fig. 1-12A. Series ohmmeter circuit.

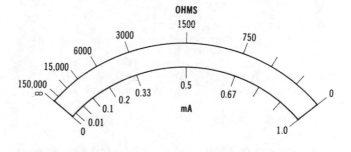

Fig. 1-12B. "Ohms" scale calibration.

one-third the full-scale current, or 0.33 mA, which means that the meter pointer will deflect to only one-third of full scale. We can then mark this point on the "ohms" scale for 3000 ohms. We can similarly calculate the amount of current and meter pointer deflection for any value of external resistance and mark the "ohms" scale accordingly. Fig. 1-12B illustrates some representative values for calibrating the "ohms" scale.

Ordinarily the usable range of a series ohmmeter is from 1/100 to 100 times the internal resistance of the circuit. Therefore there are practical limits on the amount of resistance this ohmmeter can measure. To measure lower resistances the internal resistance of the ohmmeter must be reduced. This can be done by employing a multirange ohmmeter that uses a small battery (usually 1.5 volts) for the ×1 and ×100 ranges, and a larger battery (usually 30 volts) for the ranges of ×1000 and above. To measure very low values of resistance a *shunt* ohmmeter can be used. The ohmmeter shunt is designed similarly to an ammeter shunt, except that a voltage source and an OHMS ADJUST control are placed in series with the shunt resistance.

We should keep in mind that multifunction meters are usually designed to measure voltage, current, and resistance. Most vom's, tvm's, and vtvm's used today employ one meter movement that is modified by internal circuitry to make many types of measurements. A rotary switch is ordinarily used with multifunction meters to select the proper function and range.

Objective

In this laboratory activity you will study the design of an ohmmeter. A simple ohmmeter may be built by using a basic meter movement, a current-limiting resistor, and a dc voltage source.

The value of the current-limiting resistor (R_{LIM}) is calculated in the following manner:

$$R_T = \frac{E_s}{I_M} \quad \text{and} \quad R_{LIM} = R_T - R_M$$

where
R_T is the total resistance, in ohms, required in the circuit,
E_s is the source voltage in volts,
I_M is the current in amperes, through the meter movement which causes full-scale deflection,
R_M is the resistance of the meter movement in ohms.

Equipment

Meter movement: 0-100-μA, 500-Ω, zero-centered (or equivalent)

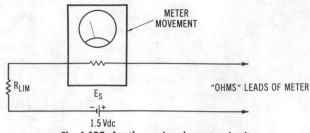

Fig. 1-12C. Another series ohmmeter circuit.

Dry cell: 1.5 V
Assorted resistors (values as calculated)

Procedure

1. Obtain a 0–100-μA meter movement and construct the circuit shown in Fig. 1-12C.
2. Calculate the value of R_{LIM} using the procedure discussed under "Objective." Insert the value of calculated R_{LIM} into the circuit. Either resistors or a potentiometer adjusted to the R_{LIM} value can be used.

   $R_{LIM} = $ _____ Ω
3. On the scale shown in Fig. 1-12D, mark "0" opposite 100 μA, and infinity (∞) opposite 0 μA.
4. Determine the resistive value (R_X) needed for half-scale deflection (50 μA). Mark this value opposite the 50-μA mark.

   $R_X = $ _____ Ω
5. Determine the current through the meter when the following values of resistance are placed across the meter leads. Record these values opposite their corresponding currents on the scale in Fig. 1-12D.

   (a) 3.75 kΩ → _____ μA

   (b) 7.50 kΩ → _____ μA

   (c) 30 kΩ → _____ μA

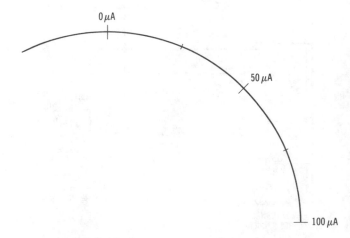

Fig. 1-12D. Microampere scale for conversion to ohms.

(d) 60 kΩ → _____ μA

(e) 120 Ω → _____ μA

6. Obtain four resistors of any value above 1 kΩ and measure them with the ohmmeter you have designed. Have your instructor approve the measurements.

(a) Resistor 1 value = _____ Ω,

meter current = _____ μA

(b) Resistor 2 value = _____ Ω,

meter current = _____ μA

(c) Resistor 3 value = _____ Ω,

meter current = _____ μA

(d) Resistor 4 value = _____ Ω,

meter current = _____ μA

Instructor's approval: _____

Analysis

1. Explain why the scale of an ohmmeter is nonlinear.

_____

_____

2. Why would a higher-voltage battery be used for an R × 10K range than a R × 1 range of a multirange ohmmeter?

_____

_____

_____

3. What is the purpose of the OHMS ADJUST control of an ohmmeter?

_____

_____

_____

4. Why should power never be applied to a circuit when resistance is being measured?

_____

_____

_____

5. Calculate the value of current-limiting resistor needed for a 0–1-mA, 200-Ω meter movement used to design a series ohmmeter circuit with a 30-V battery used as the power source.

7

AC Voltage and Current Measurements

Introduction

It is also possible to modify the basic meter movement so that it will indicate values of ac voltage or current. Half-wave and full-wave rectifier circuits can be employed to convert alternating current from an external circuit to a direct current that will produce an electromagnetic field in the coil of the meter movement. Half-wave and full-wave rectifier meter circuits are shown in Fig. 1-13A. Either of these types of rectifiers can be used to convert ac to dc to cause the pointer of the movement to deflect.

The amount of deflection of the pointer must be considered for proper scale calibration. For half-wave rectification, meter current can be expressed as:

$$I_{M(dc)} = \frac{0.45 \times V_{RMS}}{R_T}$$

where

$I_{M(dc)}$ is the dc meter current in amperes,
0.45 is a conversion factor for converting an ac rms value to an average dc value,
V_{RMS} is the maximum ac rms input voltage for the meter range in volts,
R_T is the sum of the meter resistance and the required multiplier resistance in ohms.

The ac sensitivity of a half-wave ac meter is thus only 45 percent of the dc sensitivity unless some circuit modification is made.

Full-wave rectifiers allow us to have an ac voltage range that is 90 percent of its dc equivalent. The average dc current for a full-wave rectifier can be expressed as:

$$I_{M(dc)} = \frac{0.9 \times V_{RMS}}{R_T}$$

Rectifier-type ac meters also employ blocking capacitors placed in series with the meter probes to isolate any dc component, which may be contained in the circuit being measured, from the meter circuitry. Rectifier-type meters are designed primarily for ac frequencies in the power range (60 Hz) and for audio frequencies. Other types of meters are used to measure higher frequencies.

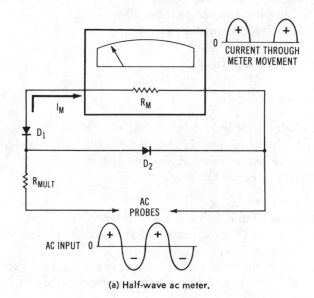

(a) Half-wave ac meter.

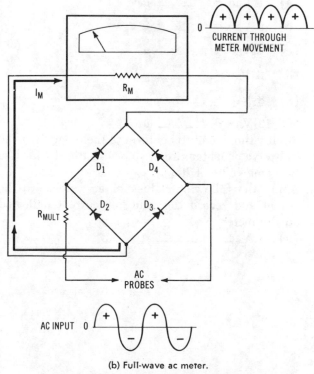

(b) Full-wave ac meter.

Fig. 1-13A. Half-wave and full-wave ac meter circuits.

43

Objective

In this activity you will use a meter movement to measure ac voltage. You will observe the process of calibrating the meter scale for ac voltage measurement.

Equipment

Meter movement: 0–100-μA, 500-Ω (or equivalent)
Variable ac power source
Multiplier resistor (as calculated)
Multifunction meter
Diodes: 1N4004 (4)
Connecting wires

Procedure

1. Refer to Fig. 1-13A. Using the meter-movement full-scale current (I_M) and resistance (R_M), determine the value of R_{MULT} needed to measure 10 V (direct current).

   $R_{MULT} = $ _____ Ω
2. Connect the half-wave ac meter circuit shown in Fig. 1-13A using two diodes and the calculated value of R_{MULT}.
3. Apply the following values of ac voltage to the circuit and record the value of current indicated on the meter:
   (a) 1 Vac → _____ μA

   (b) 2 Vac → _____ μA

   (c) 4 Vac → _____ μA

   (d) 6 Vac → _____ μA

   (e) 8 Vac → _____ μA

   (f) 10 Vac → _____ μA
4. Modify the circuit to conform to the diagram of the full-wave ac meter circuit shown in Fig. 1-13A. Use the same value of R_{MULT}.
5. Apply the following values of ac voltage to the circuit and record the value of current indicated on the meter.
   (a) 1 Vac → _____ μA

   (b) 2 Vac → _____ μA

   (c) 4 Vac → _____ μA

   (d) 6 Vac → _____ μA

   (e) 8 Vac → _____ μA

   (f) 10 Vac → _____ μA
6. This concludes the activity.

Analysis

1. Compare the values obtained in Steps 3 and 5.

   _____

   _____

   _____

2. Use the formula

$$I_{M(dc)} = \frac{0.45 \times V_{RMS}}{R_T}$$

and apply it to the half-wave circuit with 4 Vac applied (Step 3).

   $I_{M(dc)} = $ _____ μA
3. Use the formula

$$I_{M(dc)} = \frac{0.45 \times V_{RMS}}{R_T}$$

and apply it to the full-wave circuit with 4 Vac applied (Step 5).

   $I_{M(dc)} = $ _____ μA
4. Calculate the percent difference between the calculated value of $I_{M(dc)}$ in Problem 3 of the Analysis and the measured current value of the full-wave circuit (Step 5 of the Procedure).

   % Difference = _____ %

Basic Bridge Circuits

Introduction

Another group of instruments can be classified as *comparative* indicators. Usually a comparative indicator is designed to compare a known value to some unknown value. The accuracy of a comparative indicator is much better than that of a hand-deflection indicator. Meter-scale calibration is not so important for comparative instruments. Most comparative indicators are called *bridge* circuits.

Objective

This activity contains basic information about several types of bridge circuits. After you have read this information you should be familiar with bridge circuitry.

Procedure

Study the following information which pertains to bridge circuitry.

Bridge circuits are commonly used for measurement. They have high sensitivity and enable accurate measurement of resistance, capacitance, and inductance. Also, they may be used to determine impedance, reactance, frequency, and other quantities. Measurement accuracy is due largely to the null-balance method of output induction and the fact that the bridge circuit conveniently allows direct comparison of unknown components with precise standard units.

Wheatstone Bridge

The conventional Wheatstone bridge is shown in Fig. 1-14A. This bridge consists of four resistance arms, a galvanometer, and a power supply. The four arms are: (1) unknown resistance, R_X; (2) standard resistance, R_S, to which the unknown is compared; (3 and 4) two ratio arms, R_A and R_B, which provide multiplication of the range of the standard resistance in convenient values. The galvanometer (G) is used to indicate null balance. A zero-centered instrument facilitates measurements by indicating which way the

standard resistance must be adjusted to achieve a balance.

The current through the galvanometer depends upon the potential difference between points C and D.

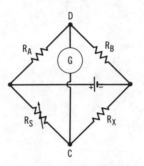

Fig. 1-14A. Wheatstone bridge.

The bridge is said to be *balanced* when there is no current through the galvanometer and the potential difference across the galvanometer is 0 volts. The balance equation for the Wheatstone bridge is:

$$R_X = R_S \times \frac{R_B}{R_A}$$

The Wheatstone bridge is widely used for precision measurement of resistance from approximately one ohm to the low megohm range.

Slide-Wire Bridge

A special form of Wheatstone bridge uses a fixed length of resistance wire instead of R_A and R_B and employs a sliding contact to establish the junction of R_A and R_B. A typical slide-wire arrangement is shown in Fig. 1-14B.

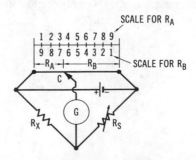

Fig. 1-14B. Slide-wire bridge.

The slide wire can be conveniently mounted upon a suitable scale, such as the one shown in Fig. 1-14B. The moving contact sets the ratio of R_A and R_B after the standard resistance (R_S) has been adjusted to one of its fixed values. The balance equation is:

$$R_X = R_S \times \frac{R_A}{R_B}$$

with R_A and R_B read directly in scale divisions. This instrument provides rapid operation and is less expensive than most bridges; however, it is less precise and has a more limited range than most bridges.

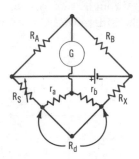

Fig. 1-14C. Kelvin bridge.

Kelvin Bridge

The Kelvin double bridge is shown in Fig. 1-14C. This bridge is a modification of the Wheatstone bridge and provides greatly increased accuracy in measurement of low-value resistances, generally below one ohm. The resistance of the leads connecting an unknown low-value resistor to the terminals of the bridge circuit may affect the accuracy of the measurement. This bridge circuit is independent of the resistance of R_d, which is usually referred to as the *yoke*. When the network resistances are so adjusted that:

$$\frac{R_A}{R_B} = \frac{r_a}{r_b}$$

the balance equation is then:

$$R_X = R_S \times \frac{R_B}{R_A}$$

For the greatest precision the ratio arms should be set at the highest practicable resistance and the con-

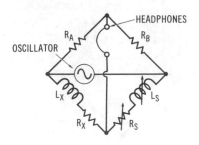

Fig. 1-14D. Resistance-ratio bridge.

nections from the bridge arms to the unknown resistance should be kept as low as possible.

The decision of when to use a Wheatstone or Kelvin bridge depends on the value of resistance to be measured and the accuracy required. With the Kelvin bridge the measurement errors due to contact resistance in the ratio arms are entirely negligible.

Resistance-Ratio Bridges

The bridges discussed previously are dc bridge circuits. Now, we should also consider ac bridge circuits. The ac bridge in its basic form consists of four bridge arms, a power supply, and a null detector. The simplest form of ac bridge is the resistance-ratio bridge, shown in Fig. 1-14D. An unknown inductance and its series resistance can be measured in terms of known inductance and resistance, which are continuously variable. For this network, the balance equations are:

$$L_X = L_S \times \frac{R_A}{R_B} \quad \text{and} \quad R_X = R_S \times \frac{R_A}{R_B}$$

where the letter symbols are as shown in Fig. 1-14D.

The resistance-ratio bridge can also be used to measure a capacitance and its series resistance in terms of a variable known capacitance and resistance. This network is shown in Fig. 1-14E. The balance equations for this circuit are:

$$C_X = C_S \times \frac{R_A}{R_B} \quad \text{and} \quad R_X = R_S \times \frac{R_A}{R_B}$$

where the letter symbols are as shown in Fig. 1-14E.

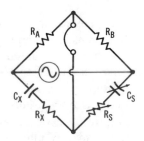

Fig. 1-14E. Capacitance bridge.

Maxwell Bridge

In the Maxwell inductance bridge, which is shown in Fig. 1-14F, an unknown inductance is balanced against a fixed standard capacitance (C_B) and two resistors (R_A and R_B) which may be adjusted for inductance balance. The principal advantage of this arrangement lies in the fact that capacitances are more nearly ideal standards than are inductance coils.

If the quality factor (Q) of the unknown inductance is not in excess of about 10 at the measurement frequency, the Maxwell bridge is well suited for the measurement of a wide range of inductions. A dis-

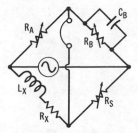

Fig. 1-14F. Maxwell bridge.

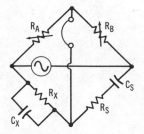

Fig. 1-14H. Wien frequency bridge.

advantage of the circuit is the interaction between reactive and resistive components when using a fixed capacitance. The balance equations for the Maxwell bridge are:

$$L_X = R_A R_S C_B \quad \text{and} \quad R_X = R_S \times \frac{R_A}{R_B}$$

$$Q = \omega \frac{L_X}{R_X} = 2\pi f C_B R_B$$

where

Q is the quality factor,
$\omega = 2\pi f$ is the angular frequency of the applied sine wave.

Hay Bridge

In the Hay bridge circuit shown in Fig. 1-14G, the standard capacitance (C_B) is in series with its associated resistances, rather than in parallel as in the Maxwell bridge. The Hay bridge is preferred for the measurement of coils having Q values of more than 10. In commercially available bridges, switching the range dial from the low-Q to high-Q setting accomplishes the internal change from the Maxwell to Hay circuits. The balance equations for the Hay bridge are very complicated and will not be considered.

Wien Bridge

The Wien bridge circuit is used to measure capacitance in terms of frequency and resistance and employs a special form of resistance-ratio bridge. It is

used for the measurement of audio frequency as well as the accurate measurement of capacitance. The Wien bridge network is shown in Fig. 1-14H. The balance equation for the Wien bridge is:

$$\frac{C_X}{C_S} = \frac{R_B}{R_A} - \frac{R_S}{R_X}$$

where the letter symbols are as shown in Fig. 1-14H.

Schering Bridge

The Schering bridge is another variation of the resistance-ratio bridge. It is used for measuring the capacitance and dissipation factor of capacitors. Shown in Fig. 1-14I, the Schering bridge measures unknown

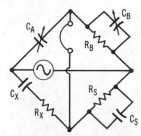

Fig. 1-14I. Schering bridge.

capacitance in direct proportion to a standard capacitance. The balance equations for this circuit are:

$$C_X = C_A \times \frac{R_B}{R_S} \quad \text{and} \quad R_X = R_S \times \frac{C_B}{C_A}$$

$$Q_X = \frac{1}{2\pi f C_X R_X}$$

where Q_X is the quality factor of the measured capacitor.

Bridges employing the Schering circuit are commercially available, particularly for the measurements of the dielectric constant, dissipation factor, and changes in characteristics with variations in frequency, temperature, and humidity.

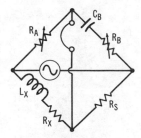

Fig. 1-14G. Hay bridge.

Wheatstone Bridge Resistance Measurement

Introduction

The Wheatstone bridge, shown in Fig. 1-15A, is a comparative instrument. The technique used for Wheatstone bridge measurement was discussed in Activity 1-14. A voltage source is used in conjunction with a sensitive zero-centered meter movement and a resistive bridge circuit. The bridge circuit is completed by adding an unknown external resistance (R_X) to be measured (see Fig. 1-15B). When resistor R_S is adjusted so that the resistive path formed by R_A and R_S is similar to the path formed by R_X and R_B, no current will flow through the galvanometer. In this condition the meter will indicate zero (referred to as a *null*) and the bridge is said to be "balanced." The value of R_S is marked on the indicator so

that the value of R_X can be determined. Resistors R_A and R_B form the "ratio arms" of the indicator. The value of the unknown resistance (R_X) is expressed mathematically as:

$$R_X = \frac{R_A}{R_B} \times R_S$$

where R_A, R_B, R_S, and R_X are arranged as shown in Fig. 1-15B.

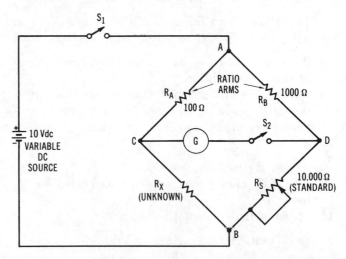

Fig. 1-15B. Wheatstone bridge circuit.

Objective

In this activity a Wheatstone bridge circuit will be constructed and tested for operation. The circuit will then be used to demonstrate how an unknown resistance is determined. A dc source is used to energize the circuit.

Equipment

Resistors: 1000 Ω, 220 Ω, 100 Ω
Potentiometers: 10,000 Ω, 1000 Ω
Galvanometer (0–100 μA)
Variable dc power source
Multifunction meter
Spst switches (2)
Connecting wires

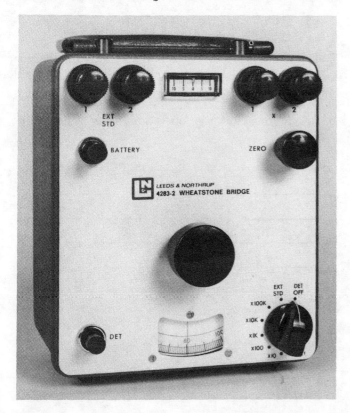

Courtesy Leeds & Northrup Co.

Fig. 1-15A. Wheatstone bridge unit.

Procedure

1. Construct the Wheatstone bridge circuit shown in Fig. 1-15B.
2. Turn on the power supply and apply 10 Vdc to the bridge.
3. Close switch S_2 to monitor the current through the galvanometer.
4. Observe and record the initial meter reading:

   _____ μAdc
5. Adjust potentiometer R_S until a null or zero-current reading is indicated on the meter.
6. Measure the following voltages with the multifunction meter:

   (a) Points A to B: _____ Vdc

   (b) Points A to C: _____ Vdc

   (c) Points B to C: _____ Vdc

   (d) Points A to D: _____ Vdc

   (e) Points B to D: _____ Vdc
7. Turn off the power source, remove R_S, and check its resistance. $R_S =$ _____ Ω (measured value) at the null condition.
8. Replace R_X with a 0–1000-Ω potentiometer. Apply power to the bridge. Record the range of the galvanometer readings as R_S is adjusted:

   $I_G =$ _____ to _____ Adc
9. Adjust R_X to midrange.
10. Adjust R_S until a null reading is indicated on the meter.
11. Measure the following voltages:

    (a) Points A to B: _____ Vdc

    (b) Points A to C: _____ Vdc

    (c) Points B to C: _____ Vdc

    (d) Points A to D: _____ Vdc

    (e) Points B to D: _____ Vdc
12. Turn off the power supply and check the resistance settings of R_X and R_S.

$R_X =$ _____ Ω

$R_S =$ _____ Ω

13. This concludes the activity.

Analysis

1. From Step 5, what is the mathematical relationship between the branches of the bridge network?

   _____

   _____

   _____

2. From Step 6, how does the voltage drop across R_S compare to the voltage drop across R_X?

   _____

3. From Step 6, how does the voltage drop across R_A compare to the voltage drop across R_B?

   _____

4. Using the resistance values (Step 12) obtained in this activity, verify that $R_X = R_S R_A / R_B$.

   _____

   _____

   _____

5. Determine the percent difference between the actual and calculated values of R_X.

   % Difference = _____ %
6. Briefly describe the operation of a Wheatstone bridge.

   _____

   _____

   _____

   _____

   _____

Balanced Wheatstone Bridge

Introduction

A balanced bridge circuit is commonly used to detect small circuit changes and to measure unknown electrical values. The bridge method of measurement reduces many of the common limitations normally associated with other test instruments. Meter scale inaccuracies, calibration errors, and circuit loading are greatly minimized with this type of measurement. The output of a bridge circuit can also be amplified to a level that ensures accurate control of practically any type of equipment or machinery. A balanced bridge will show zero or null current, while an unbalanced condition is represented by a voltage or current indication on the meter.

Objective

In this activity you will construct a balanced bridge circuit and check its operation. You will also compare balanced and unbalanced conditions.

Equipment

Dc power source
Resistors: 470 Ω (4), 100 Ω, 1 kΩ
Potentiometer: 5 kΩ
Electronic multifunction meter
Spst switch
Connecting wires

Procedure

1. Construct the balanced bridge circuit of Fig. 1-16A.
2. Close the circuit switch and prepare the meter to measure dc voltage. Measure the voltage from the source to be certain it is 5 V.
3. Measure and record the voltages across R_1, R_2, R_3, and R_4.

$V_{R1} =$ _____ Vdc

$V_{R2} =$ _____ Vdc

$V_{R3} =$ _____ Vdc

$V_{R4} =$ _____ Vdc

4. Depending upon the accuracy of the resistors, the voltages should all be equal. To test how closely the bridge circuit is balanced, connect the meter to test points A and B. An imbalance in the R_1R_3 arm or R_2R_4 arm will produce a voltage at points A and B.
5. Measure the voltage across test points A and B. If the meter deflects up scale the common and positive probe polarities are correct. If it deflects down scale the meter probe is reversed. The imbalance polarity of the bridge can also be determined by evaluating the measured voltages across each component. How does your circuit respond to this measurement?

_____

_____

6. Turn off the circuit switch and exchange the 470-Ω resistor at R_4 for a 390-Ω resistor. Turn on the circuit switch, then measure and record the voltages across R_1, R_2, R_3, and R_1.

$V_{R1} =$ _____ Vdc

$V_{R2} =$ _____ Vdc

$V_{R3} =$ _____ Vdc

$V_{R4} =$ _____ Vdc

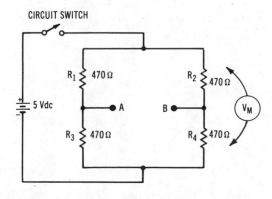

Fig. 1-16A. Balanced bridge.

7. Referring to the schematic drawing of the circuit and the measured values, which terminal (A or B) will be positive, and which will be negative (A or B)? Explain your choice.

_____

_____

_____

_____

8. This concludes the activity.

Analysis

1. What occurs in a bridge circuit when it becomes balanced?

_____

_____

_____

2. Do the voltages across each resistor in a bridge circuit have to be equal for the bridge to be properly balanced? Why?

_____

_____

_____

3. How would the polarities of the test voltages at points A and B be determined when V_{R1}, V_{R2}, V_{R3}, and V_{R4} are known?

_____

_____

_____

4. What are some advantages of using a Wheatstone bridge over the conventional ohmmeter method of resistance measurement?

_____

_____

_____

_____

AC Bridge Measurement

Introduction

An ac bridge circuit is similar in construction to a dc bridge. It generally has four circuit arms, a power source, and a null indicator. This type of bridge circuit, however, uses an ac power source such as a 60-hertz power line or an oscillator of a higher frequency. Many times the null indicator of an ac bridge is a set of headphones.

The ac bridges are used to measure unknown inductances or capacitances by comparison to a known value. Other ac circuit variables, such as power factor, total impedance, and dissipation factor can be determined by using ac bridge circuits. There are several types of ac bridge circuit modifications which are used to measure ac circuit and component values.

Objectives

In this activity you will construct a Wheatstone bridge which uses an ac voltage source. You will construct and study this ac bridge circuit.

Equipment

Electronic multifunction meter
Resistors: 100 kΩ, 200 kΩ (2)
Potentiometer: 500 kΩ
Oscilloscope
Audio signal generator or 6.3-Vac filament voltage
  source
Capacitor: 0.1 μF, 200 Vdc
Spst switch
Connecting wires

Procedure

1. Connect the ac bridge circuit of Fig. 1-17A. The ac source can be 6.3 V, 60 Hz, or from an audio signal generator.
2. Prepare the meter to measure ac voltage. Connect it to points A and B.

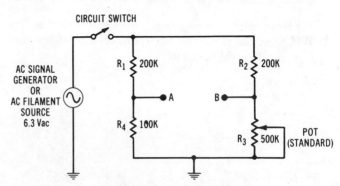

Fig. 1-17A. Ac bridge circuit.

3. Close the circuit switch. Measure and record the ac voltage across test points A and B.

    $V_{AB} =$ _____ Vac

4. Adjust the 500-kΩ potentiometer standard at R_3 to produce a zero or null reading on the voltmeter.
5. Change the voltage range to a smaller value and carefully adjust the potentiometer to a precise null.
6. Reverse the polarity of the voltmeter and test the circuit again. How does this change affect the null reading?
7. Prepare the oscilloscope for operation and connect it to the bridge circuit at test points A and B.
8. Alter the potentiometer while observing the scope. Ordinarily it is easier to null the circuit by using the oscilloscope rather than by the voltmeter method.
9. Open the circuit switch and remove R_1 and R_4. In place of R_1 connect a 0.1-μF capacitor as C_1. For R_4 place another capacitor (C_X) of any value. The oscilloscope should remain connected to points A and B.
10. Close the circuit switch and null the output of the bridge by adjusting R_3.
11. Open the circuit switch and disconnect R_3 from the bridge. With an ohmmeter measure its null setting.

    $R_3 =$ _____ Ω

12. Calculate the value of the capacitor connected in place of R_4 as C_X with the formula $C_X = C_1 R_3 / R_2$.

13. Test at least two other capacitors to check the accuracy of this test procedure.
14. This concludes the activity.

Analysis

1. Why must ac be used to determine the value of a capacitor?

_____

_____

_____

2. What factors limit the accuracy of the bridge circuit used in this activity?

_____

_____

_____

3. How could an unknown inductance be measured using an ac bridge?

_____

_____

_____

_____

Electronic Bridge

Introduction

A recent innovation in electronic bridge circuitry is the operational-amplifier (op-amp) photoelectric bridge circuit. In this type of circuit a photoconductive transducer is connected to one arm of a bridge circuit and the output is connected to an operational amplifier.

The operational amplifier is used to amplify changes in light intensity. This type of circuit is only one example of comparative bridge circuits which could be used for measurement. Op amps are easily used in such circuits to amplify small changes of light intensity.

Objective

In this activity you will construct an op-amp bridge controlled by a photoelectric circuit. This circuit is used to cause variation of an output indicator. The indicator could be calibrated to measure changes of light intensity.

Equipment

Dc power supply
Electronic multifunction meter
Photoconductive cell (GE-X6 or equivalent)
Operational amplifier (LM 3900 or equivalent)
Resistors: 100 kΩ (2), 10 kΩ, 470 Ω
Potentiometer: 500 kΩ
60-W lamp with socket
Variable ac power supply
Spst switch
Connecting wires

Procedure

1. Construct the photoelectric bridge circuit of Fig. 1-18A.

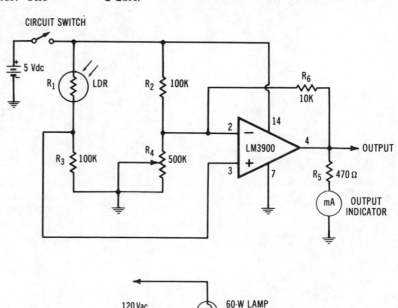

Fig. 1-18A. Electronic bridge circuit.

2. Prepare a 3-inch-long paper tube and attach it to the photoconductive cell. Position the tube and the photocell so that the tube points directly to the light source.

3. Close the circuit switch and adjust resistor R_4. You should be able to block the light source 12 inches or more away from the paper tube and cause the output indicator to change.

4. Prepare an electronic multifunction meter to measure dc voltage and connect it from ground to terminal 3 of the op amp.

5. Block the light from the source so that it cannot reach the cell. Is the bridge balanced or unbalanced?

6. Allow light to shine onto the cell. How does the bridge respond?

7. The total change in input voltage between the balanced and unbalanced condition of the bridge is from _____ V to _____ V.

8. Measure the output voltage of the op amp. Test the total change in output voltage when light is shined on the circuit and when light is blocked.

The voltage change is from _____ to _____ V.

9. Calculate the voltage gain of the op-amp bridge circuit as follows:

$$\frac{V_{out}}{V_{in}} = \underline{\hspace{2cm}}$$

10. Check the change in output current caused by light-intensity changes. The output current changes from _____ mA to _____ mA.

11. This concludes the activity.

Analysis

1. What is the primary function of the op amp in Fig. 1-18A?

_____

_____

_____

2. What is the equivalent resistance-ratio formula of the bridge circuit when it is balanced?

_____

_____

_____

3. What is the function of R_6 in Fig. 1-18A?

_____

_____

_____

4. What would be some possible applications of this circuit?

_____

_____

_____

_____

_____

Potentiometric Measurement

Introduction

Another type of comparative instrument is the *potentiometer*. While the bridge method is used to measure impedance, the potentiometric method is used to measure voltages accurately. The illustration of Fig. 1-19A shows the simplified potentiometer method. The voltage to be measured is applied to the input terminals of the circuit. The potentiometer, which is connected across the reference supply voltage, forms a voltage divider. When the movable arm of the potentiometer is adjusted to a voltage (V_{REF}) equal to the voltage being measured, the current through the meter will be zero. When the zero (or *null*) reading on the zero-centered meter is obtained, the value of the unknown voltage can be read from a calibrated scale on the potentiometer.

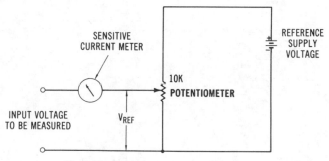

Fig. 1-19A. Potentiometer comparison circuit.

Since no current is drawn from the voltage source being measured during the null condition, the meter can be considered an infinite impedance. Thus the potentiometric technique can be used for precise calibration of other instruments. Potentiometric instruments can also be used as laboratory voltage standards. Potentiometers either use a standard cell (see Fig. 1-19B), which has a precise voltage, or a zener-diode standard reference voltage supply.

Objective

In this activity you will construct and test a simplified potentiometric comparison circuit.

Courtesy Leeds & Northrup Co.

Fig. 1-19B. Standard cell.

Equipment

Variable dc power source
Zero-centered current meter
Dry cell: 1.5 V (3)
Potentiometer: 10 kΩ

Procedure

1. Construct the circuit shown in Fig. 1-19A. Use a 10-kΩ potentiometer. Adjust the power supply to 1.5 Vdc and use this as the reference supply voltage. The input voltage to be measured should be from a 1.5-V dry cell. Use a zero-centered meter as the sensitive current meter.
2. Adjust the potentiometer until a null reading is obtained. In a commercial potentiometer instrument the voltage of the unknown input voltage (dry

cell) would be indicated on a calibrated scale as the shaft of the potentiometer is rotated.

3. Remove the dry cell and replace it with another dry cell. Record the current indicated by the meter.

   Current = _____

4. Adjust the potentiometer until a null condition is obtained.

5. Remove the dry cell and replace it with another dry cell. Record the current indicated on the meter.

   Current = _____

6. Slowly increase the reference supply voltage. What occurs?

   _____

7. This concludes the activity.

Analysis

1. What does the difference in current readings in Steps 3 and 5 indicate?

   _____

   _____

   _____

2. Why is a standard cell used as a reference voltage source?

   _____

   _____

   _____

3. Briefly discuss a potentiometric comparison circuit.

   _____

   _____

   _____

   _____

Frequency Measurement

Introduction

Another very important measurement is frequency. Frequency refers to the number of cycles of voltage or current that occur in a given period. The international unit of measurement for frequency is the hertz (Hz), which means one cycle per second. A table of frequency bands is shown in Fig. 1-20A. The standard power frequency in the United States is 60 Hz. There are many frequency ranges, as shown in the figure.

Frequency can be measured with several different types of instruments. An electronic counter is one type of frequency indicator. Vibrating-reed frequency indicators are common for measuring power frequencies. An oscilloscope can also be used to measure frequency. Graph-recording instruments can also be used to provide a visual display of frequency.

One common method of frequency measurement for sine-wave voltages is a comparative method which uses the oscilloscope. This technique relies upon so-called Lissajous patterns on the oscilloscope screen. An unknown frequency may be applied to the vertical input of the oscilloscope while a known value of frequency is applied to the horizontal input. The shape of the pattern which appears on the screen of the oscilloscope is used to determine the ratio of the vertical frequency to the horizontal frequency. When this ratio has been determined, the unknown frequency can be easily calculated.

The formula used to find the unknown frequency value is as follows:

$$f_x = f \times \frac{T_H}{T_V}$$

where

f_x is the unknown frequency,
f is the known frequency,
T_H is the number of times the Lissajous pattern touches a horizontal line on the oscilloscope screen,
T_V is the number of times the pattern touches a vertical line on the screen.

Some examples are shown in Fig. 1-20B. Note that one horizontal line and one vertical line are drawn tangent to the pattern.

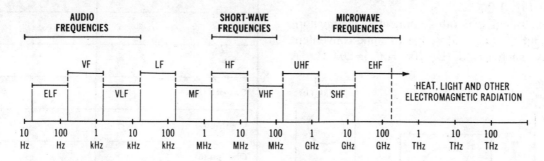

BAND		FREQUENCY RANGE		
EXTREMELY LOW FREQUENCY	(ELF)	30Hz	TO	300Hz
VOICE FREQUENCY	(VF)	300Hz	TO	3kHz
VERY-LOW FREQUENCY	(VLF)	3kHz	TO	30kHz
LOW FREQUENCY	(LF)	30kHz	TO	300kHz
MEDIUM FREQUENCY	(MF)	300kHz	TO	3MHz
HIGH FREQUENCY	(HF)	3MHz	TO	30MHz
VERY-HIGH FREQUENCY	(VHF)	30MHz	TO	300MHz
ULTRA HIGH FREQUENCY	(UHF)	300MHz	TO	3GHz
SUPER-HIGH FREQUENCY	(SHF)	3GHz	TO	30GHz
EXTREMELY HIGH FREQUENCY	(EHF)	30GHz	TO	300GHz

Fig. 1-20A. Classification of frequency bands.

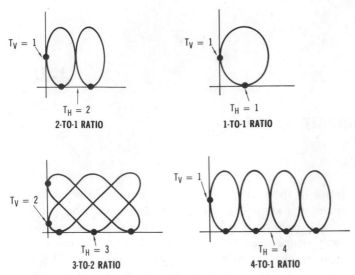

Fig. 1-20B. Lissajous patterns.

low-voltage 60-Hz source can also be used as the horizontal input.

3. Record the value of frequency to which the signal source is set.
   Known frequency = _____ Hz

4. Connect an audio signal generator to the vertical input of the oscilloscope. Use this as the "unknown" frequency.

5. Adjust the sweep select control of the oscilloscope to the EXTERNAL setting.

6. Adjust the signal generator to a frequency which is equal to that of the known frequency.

7. Make a sketch of the pattern which appears.

Objective

In this activity you will use the oscilloscope to produce Lissajous patterns to determine unknown frequencies.

Equipment

Oscilloscope
Audio signal generators (2) or 1 audio signal generator and low-voltage 60-Hz ac source
Connecting wires

Procedure

1. Set up an oscilloscope and signal generator as shown in Fig. 1-20C.

2. Adjust the signal generator connected to the horizontal input of the oscilloscope to some convenient frequency, such as 60 Hz, 100 Hz, or 200 Hz. A

8. Adjust the signal generator to the following multiples of the known frequency and make sketches of the patterns which appear on the oscilloscope in Table 1-20A.

Table 1-20A. Oscilloscope Patterns

Multiple of Known Frequency	Oscilloscope Pattern
Two	
Three	
Four	
One-Half	
One-Third	
One-Fourth	

9. This concludes the activity.

Analysis

1. Using the formula $f_x = f\, T_H / T_V$ determine the unknown frequencies associated with the following patterns (assume $f = 500$ Hz):

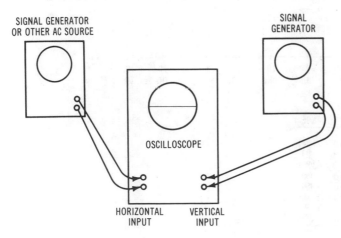

Fig. 1-20C. Experimental setup for frequency comparison.

(a)

(b)

(c)

(d)

(e)

(f)

2. What are some other methods of measuring frequency?

_____

_____

_____

_____

_____

_____

_____

Numerical Readout Instruments and Readout Devices

Introduction

Many instruments now in use employ numerical readouts to simplify the measurement process and to make more accurate measurements. Instruments such as digital counters, digital multimeters, and digital voltmeters are commonly employed for measurement. Numerical readout instruments, such as the one shown in Fig. 1-21A, rely upon the operation of gating and flip-flop circuitry in order to produce a digital readout of the measured quantity.

The readout of a digital instrument is primarily designed to transform electrical signals into numerical presentations. Both letter and number readouts are available. Seven-segment, discrete-number, and bar-matrix displays are currently available. Each method has a unique device that is designed to change electrical energy into light energy. The basic characteristics of the device dictates such things as operating voltage, current, illumination level, and quality of the display character.

The display of a digital system is achieved by several distinct electronic processes. These include the ionization of gas, heated incandescent elements, and light-emitting diodes.

Objective

This activity will investigate one of the types of numerical readout devices which are used with electronic instruments. A Nixie tube has ten cathode elements which glow in the shape of numbers from zero (0) through nine (9). Nixie tubes were one of the first types of numerical readout devices used. The term "Nixie" is a trademark of the Burroughs Corporation; however, other manufacturers supply similar readout devices.

The major operational part of a Nixie tube is the cathode sections. There are ten separate cathodes inside the tube which glow with an orange color when they become ionized by a voltage input. The Nixie tube has one anode and ten cathodes as shown in Fig. 1-21B. When this type of tube is used as a digital readout, a cathode is energized by a voltage input which causes neon gas surrounding the cathode to ionize. The cathodes are stacked side-by-side inside the tube, but the glow of the neon gas is localized to

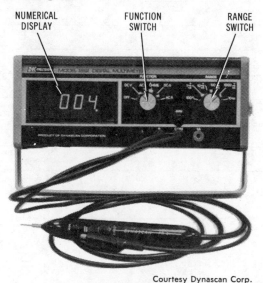

NUMERICAL DISPLAY   FUNCTION SWITCH   RANGE SWITCH

Courtesy Dynascan Corp.

Fig. 1-21A. Digital readout meter.

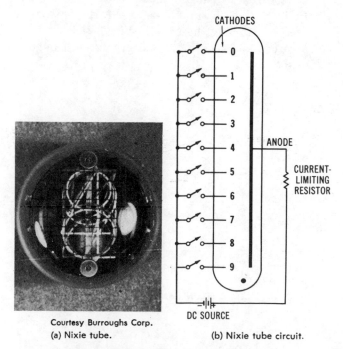

Courtesy Burroughs Corp.
(a) Nixie tube.   (b) Nixie tube circuit.

Fig. 1-21B. Nixie tube and circuit diagram.

the cathode when it is energized. Therefore only one number is displayed on the readout unless the decoder-driver circuitry has a malfunction.

Equipment

Nixie tube with socket
Dc power supply
Spst switch
Electronic multifunction meter
Resistor: 10 kΩ
Connecting wires

Procedure

2. Connect the Nixie tube readout circuit of Fig. 1-21C. There is such a wide range of base connections used with this type of display only the display numbers are used as reference. Specific device base wiring diagrams are supplied by the manufacturer.
2. Determine the anode of the tube. Ordinarily it can be easily determined because it is a platelike structure of screen mesh. The positive side of the dc power supply must be attached to this element.
3. Specific number pin connections usually can be determined by their internal connection to the display number.
4. Turn on the circuit switch and adjust the power supply to 170 Vdc. Connect the negative terminal of the power supply alternately to each pin connection other than the anode lead.
5. Indicate the pin numbers of the display device used by labeling the tube in Fig. 1-21B.
6. Measure and record the current to each number of the display.

0 = _____ mA

1 = _____ mA

2 = _____ mA

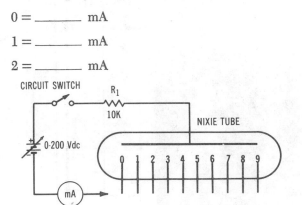

Fig. 1-21C. Nixie tube test circuit.

3 = _____ mA

4 = _____ mA

5 = _____ mA

6 = _____ mA

7 = _____ mA

8 = _____ mA

9 = _____ mA

7. Prepare the display to indicate the number 8.
8. Carefully decrease the supply voltage until deionization occurs. The deionization occurs at _____ Vdc.
9. After the tube becomes deionized, increase the supply voltage until the tube illuminates. This is called the *ionization potential*. The ionization potential is _____ Vdc.
10. Do not attempt to reverse the polarity of the source, because the device could be permanently damaged.
11. This concludes the activity.

Analysis

1. Discuss gaseous-tube readout devices.

_____

_____

_____

_____

_____

_____

_____

_____

_____

_____

_____

_____

_____

Incandescent Instrument Readout Devices

Introduction

Incandescent readout devices employ seven discrete resistive bar elements suspended between supporting posts. Illumination occurs when current flows through a specific element and a common point. Approximately 5 V of ac or dc is needed to provide illumination.

Readout devices of this type are commonly called *Numitrons*. RCA is the principal manufacturer of this device.

Fig. 1-22A shows a picture of a DR2000 Numitron and its electrical circuit. Note that this readout produces a block type of seven-segment numbers similar to that of the gaseous readout device. The chief advantage of the Numitron is its variable intensity characteristic. The filament segments of this device are, however, somewhat fragile when they are energized.

When a segment of a Numitron is electrically energized, heat occurs. If enough heat is developed, the filament wire changes to a dull orange appearance and produces light energy. The degree of illumination produced is quite evident when compared with an unenergized element. As a result this noticeable change is used to indicate segment illumination.

The filament segments of a Numitron are connected together to form a single common point. This means that each filament presents a parallel path for the current from the source. As a result of this construction, filament current increases a set amount when each segment is energized. The number 8 demands the largest amount of current from the source when it is displayed.

The circuitry of a Numitron is somewhat simplified when compared with other seven-segment readout devices. It does not require a current-limiting resistor. This resistance is self-contained in each filament element. The common point is typically connected to the positive side of the source, and each segment is energized by connecting its other side to ground.

Objective

In this activity you will observe the operation of an incandescent readout device.

(a) Photograph.

Courtesy RCA

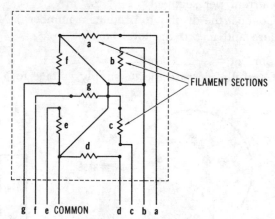

FILAMENT SECTIONS

g f e COMMON   d c b a

(b) Electrical diagram.

Fig. 1-22A. Numitron readout device.

Equipment

Dc power supply
Electronic multifunction meter
Incandescent readout device: RCA DR2000 or equivalent seven-segment readout
Connecting wires

Procedure

1. Connect the incandescent seven-segment readout circuit of Fig. 1-22B. Either a 16-pin dual in-line package (DIP) or a 9-pin tube base type of display may be used.
2. If the 9-pin base type of display is used, connect the positive side of the 5-V source to the common lead at pin 2. Turn on the dc power supply and determine the segment pin numbers by touching the negative lead alternately to each pin except 2. Indicate the corresponding pin number for each segment below.

   a = _____   e = _____

   b = _____   f = _____

   c = _____   g = _____

   d = _____

3. If the DIP type of readout is used, connect pins 2, 5, 10, 12, and 13 together. This common point is then connected in series with a milliammeter to the positive side of the 5-V dc power supply. Turn on the power supply and determine the segment pin numbers by touching the negative lead alternately to each pin except 2, 5, 10, 12, and 13. Record the corresponding pin number for each segment in the space above.
4. Record the current that occurs when each segment is illuminated.
   Current per segment = _____ mA
5. Connect the display to produce a number 8. Measure and record the total current.

   Current = _____ mA
6. Carefully increase the dc supply voltage to 5.5 V and 6 V. How does this alter the operation of the readout?

   _____

   _____

   _____

7. Lower the dc supply voltage to 4.5 V, 4.0 V, and 3.5 V. How does this alter the operation of the readout?

   _____

   _____

   _____

8. Turn off the dc power supply and remove the milliammeter from the circuit. Reverse the polarity of the dc power supply using 5 Vdc. Turn on the power supply and test the circuit operation. How does this alter the operation of the readout?

   _____

   _____

   _____

9. Turn off the dc power supply and connect a 6.3-Vac source to the display. Turn on the circuit and observe the display. How does this change the operation of the readout?

   _____

   _____

   _____

10. This concludes the activity.

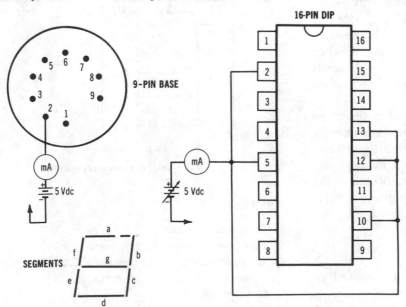

Fig. 1-22B. Incandescent readout.

Analysis

1. Prepare a chart showing the pin connections needed to display the numbers 0–9 on an incandescent display.

2. What are some of the advantages of an incandescent type of display?

_____

_____

_____

_____

_____

_____

_____

Light-Emitting Diode (LED) Readout Devices

Introduction

Light-emitting diodes are commonly used in seven-segment and 5×7 dot matrix readouts. The LEDs of these devices produce visible light when forward biased, and no light when reverse biased. As a result of this two-state condition, individual segments or dots can be illuminated when diodes are energized. Typically the positive side of the energy source is connected to the anodes of each diode through a current-limiting resistor. The cathode of a diode is then grounded. When the circuit is completed, the diode is energized and produces light.

Seven-segment LED readout devices often contain four or more LEDs connected in parallel to form a segment. This type of construction usually necessitates only one current-limiting resistor for each segment. The amount of voltage needed to produce illumination is typically 3.5 to 5 Vdc.

Fig. 1-23A shows the circuitry of seven-segment and 5×7 dot matrix LED readout devices. The LEDs in the two circuits are similar in all respects. The switching method needed to energize specific diodes is somewhat different. In the seven-segment device each segment is controlled by a single switch. The dot matrix circuit is controlled by two or more switches. An LED can be energized by two switches, such as row 4, column 5. A complete vertical row would require one column switch and all seven row switches. A complete horizontal row would be energized by one row switch and all five column switches. Dot matrix readout devices are used to produce letter displays rather than numbers. LED readout devices are used in many types of instruments.

Objective

In this activity you will observe the operation of a light-emitting diode (LED) seven-segment readout device. These devices are commonly used in instruments today.

Equipment

Dc power source
Electronic multifunction meter
LED readout device: MAN-1
Resistors: 470 Ω (8); 390 Ω
Connecting wires

Procedure

1. Connect the LED seven-segment readout circuits of Fig. 1-23B.
2. Turn on the power supply and adjust it to 5 Vdc. Then turn on the circuit switch.
3. Alternately connect each of the 470-Ω resistors to the ground or common side of the power supply. Label the specific pin number that causes illumination of each segment of the readout.

   a = _____; b = _____; c = _____;

   d = _____; e = _____; f = _____;

   g = _____.
4. The dc current needed to illuminate one segment

   is _____ mA.
5. Connect the resistor combination needed to produce a number 8. The current needed to produce

   illumination of all seven segments is _____ mA.
6. Open the circuit switch and reverse the polarity of the three common leads, 3, 9, and 14, so that they are now attached to the negative side of the power supply. Close the circuit switch and alternately connect each resistor to the positive side of the dc supply. How does this operation alter the display?

   _____

   _____

   _____

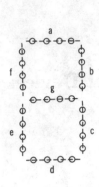

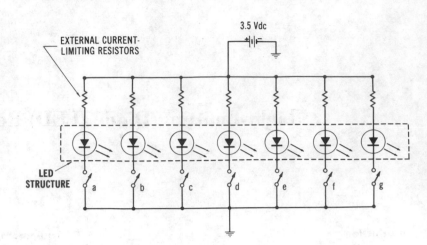

(a) 7-segment LED display with 4 diodes per segment.

(b) Diagram showing connections of segments.

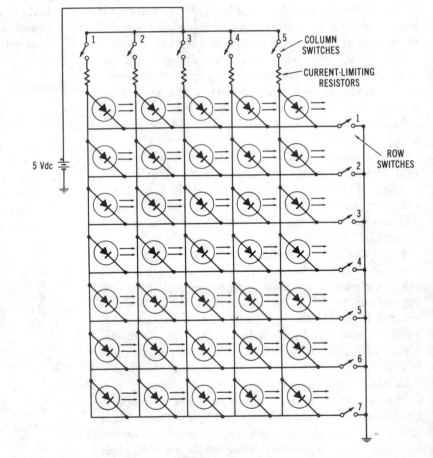

(c) 5 × 7 LED dot matrix display.

(d) 5 × 7 LED dot matrix diagram.

Fig. 1-23A. LED readout devices.

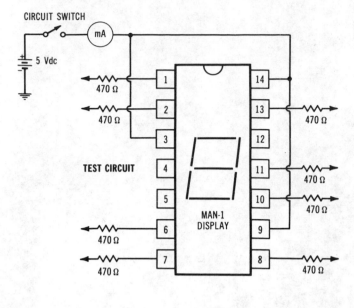

TEST CIRCUIT

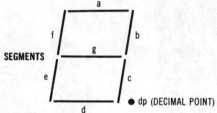

Fig. 1-23B. LED readout test circuit.

7. Open the circuit switch and return the power supply polarity to its original condition. Remove the milliammeter from the circuit and connect the common leads to the positive side of the dc power source.

8. Connect segment *f* so that it produces illumination when energized. Remove the 470-Ω resistor from segment *e* and place a 390-Ω resistor in its place. Connect the milliammeter in series with the 390-Ω resistor.

9. Close the circuit switch and observe the level of illumination produced by segments *f* and *e*. The current in segment *e* is _____ mA. What influence does the increased current have upon segment illumination?

_____

_____

10. This concludes the activity.

Analysis

1. How does the segment current of an LED readout compare with that of the incandescent readout of Activity 1-22?

_____

_____

2. How do the source polarities of an LED readout and the incandescent readout compare?

_____

_____

3. How do the illumination levels of the LED and incandescent readouts compare?

_____

_____

4. What are some of the unique differences in the gaseous-device, LED, and incandescent readouts?

_____

_____

_____

_____

Thermal Input Transducers for Instrumentation Systems

One of the most measured quantities in industry today is temperature. Without accurate information relative to the temperature of the materials used in certain manufacturing processes, as well as information associated with the manufacturing environment, it would be impossible to produce thousands of the products available to the consumer today.

Temperature may be measured by a variety of methods, most of which employ temperature transducers. Some of the more common temperature transducers used to measure temperature in an industrial environment are thermocouples, thermistors, and rtd's (resistance temperature detectors).

Thermocouples are generally thought of as being two dissimilar conductors joined at one end. The junction formed where the two conductors are joined is the measurement junction and is placed in the environment where the temperature is to be measured. As the measurement junction is heated, a millivolt output appears at the cool, unconnected ends of the thermocouple. This millivolt output is proportional to and indicative of the temperature of the measurement junction.

Thermistors are special temperature-sensitive devices formed from various oxides into discs, beads, rods, and washers. Actually, these components are thermally sensitive resistors that exhibit a negative temperature coefficient. As the temperature of the environment of a thermistor increases, the resistance of the thermistor decreases.

Resistance temperature detectors (rtd's) are conductive devices formed from materials such as plat-

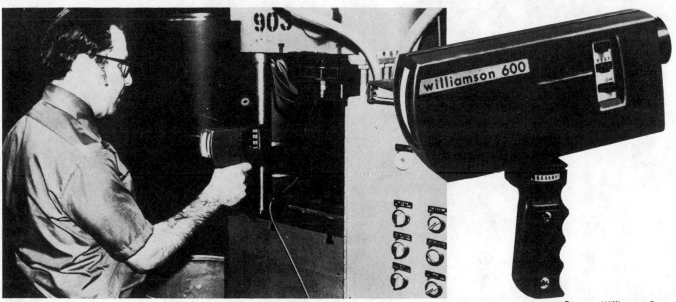

Courtesy Williamson Corp.

Indirect temperature measurement.

73

inum, nickel, and tungsten. These devices exhibit a positive temperature coefficient. This means that as the temperature of an rtd increases, its resistance likewise increases.

All of the transducers described thus far change a temperature to an electrical signal either directly or indirectly. Many of them employ amplifiers to increase the amplitude of their outputs.

In this unit you will have an opportunity to examine the characteristics and applications of the thermocouple, thermistor, and resistance temperature detector. You will also become familiar with temperature measurement and conversion using the Fahrenheit, Celsius, Kelvin, and Rankine scales. Finally, you will make certain mathematical computations pertinent to measuring temperature.

Hand-held "temperature viewer."

Courtesy Williamson Corp.

Temperature Measurement and Conversion

Introduction

The measurement and control of temperature has been of considerable human concern since our ancestors' first attempts to force crude tools from metal. With progress and experimentation, we have found that the more accurately we can measure temperature, the better we can control both it and the manufacturing process. The accurate measurement of temperature is still very important in industry. Of all of the major variables to be measured and controlled in an industrial setting, well over half are related to the measurement of temperature.

Courtesy Mikron Instrument Co.

Hand-held type infrared thermometer.

Through experimentation to measure temperature accurately, a series of scales has been devised. As the result of many individuals' experimentation and efforts, there are four widely recognized temperature measurement scales in use today. These are the Fahrenheit, Celsius (or centigrade), Kelvin, and Rankine scales. Fig. 2-1A compares these temperature scales.

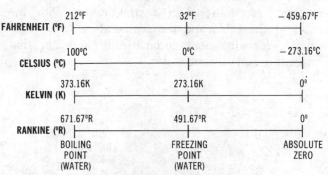

Fig. 2-1A. Prevailing temperature scales.

Since these scales are widely used in industrial settings to measure temperature, it is frequently advantageous to change the temperature readings of one scale to another. Because all of these scales are related, conversion from one scale to another may be accomplished mathematically by applying the proper conversion formula.

For example, to convert degrees Fahrenheit (°F) to degrees Celsius (°C), one of the following formulas is used:

$$°C = \frac{5}{9}(°F - 32°) \quad \text{or} \quad °C = \frac{°F - 32°}{1.8}$$

To change degrees Celsius to degrees Fahrenheit requires the use of one of the following formulas:

$$°F = \frac{9}{5}(°C) + 32° \quad \text{or} \quad °F = 1.8(°C) + 32°$$

Degrees Celsius may be converted to kelvins (degrees Kelvin) by using this formula:

$$K = °C + 273.16°$$

Likewise, kelvins can be converted to degrees Celsius by simply changing the same formula to become:

$$°C = K - 273.16°$$

where K is the number of kelvins.

Degrees Fahrenheit may be changed to degrees Rankine (°R) by using this formula:

$$°R = °F + 459.67°$$

Finally, degrees Rankine can be converted to degrees Fahrenheit by changing the preceding formula to become:

$$°F = °R - 459.67°$$

A combination of these formulas enables temperatures from any scale to be changed to any of the three remaining scales.

Changing temperatures from one scale to another may also be accomplished by using a temperature conversion chart similar to Fig. 2-1B. To use this chart correctly, one must align a straightedge with the center of the table and the temperature to be converted. This alignment automatically indicates that temperature on all four scales.

Objective

In this activity you will measure the temperature of ice and water and convert these measurements to the other units of temperature. You will also use a temperature conversion chart and temperature conversion formulas to convert some preselected values.

Equipment

Bimetal thermometer (0° to 250°F or −17.7°C to 121.1°C)
Styrofoam cup
Ice
Hot tapwater

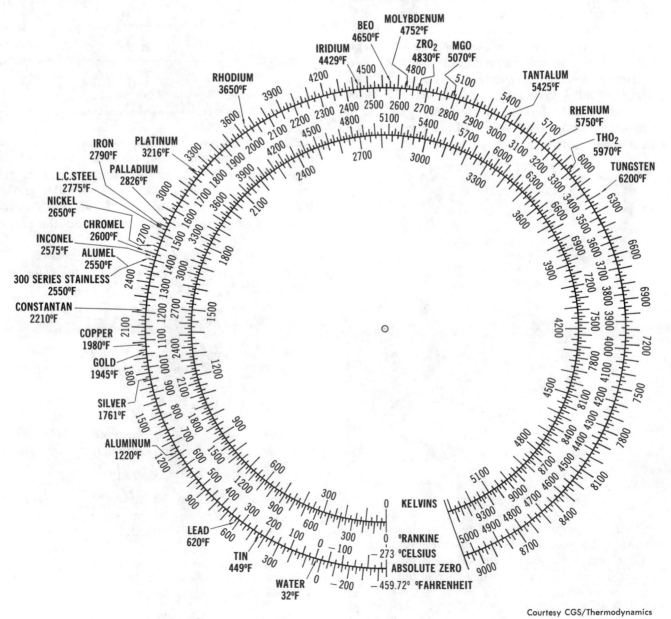

Courtesy CGS/Thermodynamics

Fig. 2-1B. Temperature conversion chart showing melting points of commonly used elements.

Courtesy Mikron Instrument Co.

Infrared thermometer.

Procedure

1. Grasp the sensing end of the thermometer in your hand for five (5) minutes.
2. Record the temperature indicated by the thermometer on the following table. Complete this table by converting the degrees Fahrenheit to degrees Celsius, kelvins, and degrees Rankine.

°F	°C	K	°R

3. Place ice in the styrofoam cup and insert the sensing end of the thermometer into the ice for about five (5) minutes.
4. Complete the following table by recording the temperature of the ice in degrees Fahrenheit and converting this reading to degrees Celsius, kelvins, and degrees Rankine.

°F	°C	K	°R

5. Replace the ice in the cup with hot tapwater and insert the sensing end of the thermometer into the water for about five (5) minutes.
6. Complete the following table by recording the temperature of the water in degrees Fahrenheit and converting this reading to degrees Celsius, kelvins, and degrees Rankine.

°F	°C	K	°R

Analysis

1. Why is it frequently necessary to change temperatures from one scale to another?

_____

_____

_____

2. What two methods may be used to convert temperatures from one scale to another?

_____

_____

_____

_____

3. Using the temperature conversion chart included as a part of this activity, convert the following temperatures from degrees Fahrenheit to degrees Celsius, kelvins, and degrees Rankine.

°F	°C	K	°R
8100			
6000			
4500			
2700			
1500			
300			

4. Using the formulas included in the introduction of this activity, change the following temperatures from degrees Fahrenheit to degrees Celsius, kelvins, and degrees Rankine.

°F	°C	K	°R
500			
1100			
−200			
−350			
3900			
5700			

5. Using either method of conversion, change the following temperatures to the indicated scale.

600°R = _____ °C

1200 K = _____ °F

1000°C = _____ °R   3300 K = _____ °R

9000°R = _____ K   −273°C = _____ K

Seebeck and Peltier Thermal Effects

Introduction

Many scientists have experimented with the thermal phenomenon throughout history. The most notable of these were T. J. Seebeck and Jean C. A. Peltier for whom the Seebeck effect and the Peltier effect were named.

The Seebeck effect is demonstrated when two dissimilar conductors are fastened together at both ends and one of the junctions is heated. This will cause an electric current to flow in the circuit formed by the joined conductors as long as there is a difference in the temperatures of their junctions. The greater the difference in the junction temperatures, the greater the current in the circuit.

The Peltier effect is shown when current is caused to flow through a junction formed where two pieces of dissimilar metals are joined. The junction through which current is passing will be either heated or cooled, depending upon the direction of the current. The greater the current, the greater the heating or cooling of the junction.

The operation of thermocouple sensors or transducers depends upon the Seebeck effect to measure temperature.

Objective

In this activity you will observe the Seebeck effect for different temperatures.

Equipment

Type J thermocouples (2)
Zero-centered 50-μA meter
660-W heat cone
120-Vac power supply
Connecting wires

Procedure

1. Construct the circuit as illustrated in Fig. 2-2A using two type J thermocouples.

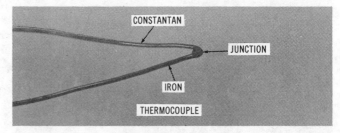

Thermocouple (bare wire).

2. Position junctions 1 and 2 to cause each to be exposed to the identical room temperature.
3. Record the current indicated by the microammeter with both junctions at the same temperature.

   $I = $ _____ μA

4. Grasp junction 1 with your hand for a period of 1 minute. Record the amount of current caused by this action as indicated by the microammeter.

   $I = $ _____ μA

5. Indicate the direction of meter deflection when junction 1 is grasped. (Check one.)

   ☐ From the center to the right.

   ☐ From the center to the left.

6. Grasp junction 2 with your hand for a period of 1 minute. Record the amount of current caused by this action as indicated by the microammeter.

   $I = $ _____ μA

7. Indicate the direction of meter deflection when junction 2 is grasped. (Check one.)

   ☐ From the center to the right.

   ☐ From the center to the left.

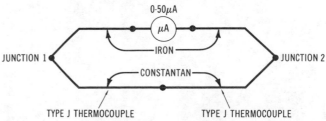

Fig. 2-2A. Thermocouple circuit.

8. How does the current recorded in Step 4 compare with that recorded in Step 6?

_____

_____

_____

9. How does the direction of meter deflection observed in Step 5 compare with the deflection in Step 7?

_____

10. Why does the microammeter deflect in opposite directions when junction 1 and junction 2 are grasped?

_____

_____

_____

11. Connect 120 Vac to the 660-watt heat cone and allow it to "warm up" for about 5 minutes. (NOTE: This cone is *very hot*. Extreme caution should be exercised while working with this device.)

12. Place junction 1 within 1 inch (1.27 cm) of the heat cone for a period of 1 minute. Record the current caused by this extreme temperature.

   $I =$ _____ μA

13. Place junction 2 within 1 inch (2.54 cm) of the heat cone for a period of 1 minute. Record the current caused due to this high temperature.

   $I =$ _____ μA

14. How does the current in Step 12 compare with the current in Step 13?

_____

_____

Analysis

1. Define and discuss the Seebeck effect.

_____

_____

2. Define and discuss the Peltier effect.

_____

_____

_____

3. When the temperature of one of the junctions in the circuit of Fig 2-2A is increased, why does current flow?

_____

_____

_____

4. How much current will flow in the illustrated circuit, if both junctions are at the same temperature?

_____

5. What is the relationship between difference in junction temperatures and current in the illustrated circuit?

_____

_____

_____

6. Where is the Seebeck effect used in temperature measurement?

_____

_____

_____

_____

_____

Thermocouple Characteristics

Introduction

When certain types of dissimilar metals are joined at one end, a thermocouple is formed (Fig. 2-3A). When the joint or junction is heated, an emf in millivolts is created between the cool ends of the dissimilar metals. The millivolt output at the cool ends of the metals is indicative of the temperature at the heated junction.

The thermocouple is the most frequently used device for measuring temperature. The Instrument Society of America has identified several standard thermocouple metal combinations currently used in industry to measure temperature. Some of the more common metal combinations are identified by their ISA code in Table 2-3A.

All of the above thermocouples have certain advantages and disadvantages. The Type J thermocouple is

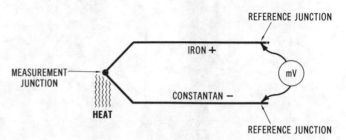

Fig. 2-3A. Type J thermocouple.

used frequently in industry because it is economical and has a relatively high millivolt output over its temperature range.

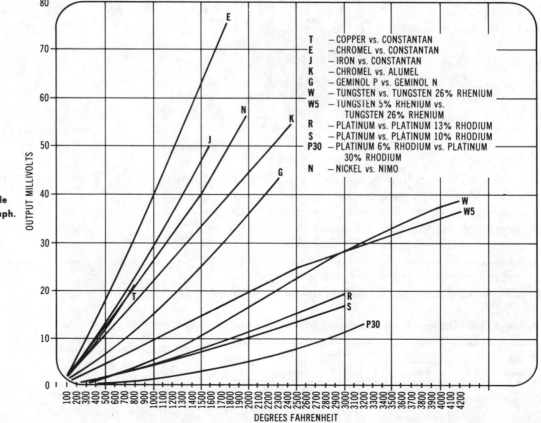

Fig. 2-3B. Thermocouple temperature-millivolt graph.

T — COPPER vs. CONSTANTAN
E — CHROMEL vs. CONSTANTAN
J — IRON vs. CONSTANTAN
K — CHROMEL vs. ALUMEL
G — GEMINOL P vs. GEMINOL N
W — TUNGSTEN vs. TUNGSTEN 26% RHENIUM
W5 — TUNGSTEN 5% RHENIUM vs. TUNGSTEN 26% RHENIUM
R — PLATINUM vs. PLATINUM 13% RHODIUM
S — PLATINUM vs. PLATINUM 10% RHODIUM
P30 — PLATINUM 6% RHODIUM vs. PLATINUM 30% RHODIUM
N — NICKEL vs. NIMO

Courtesy CGS/Thermodynamics

81

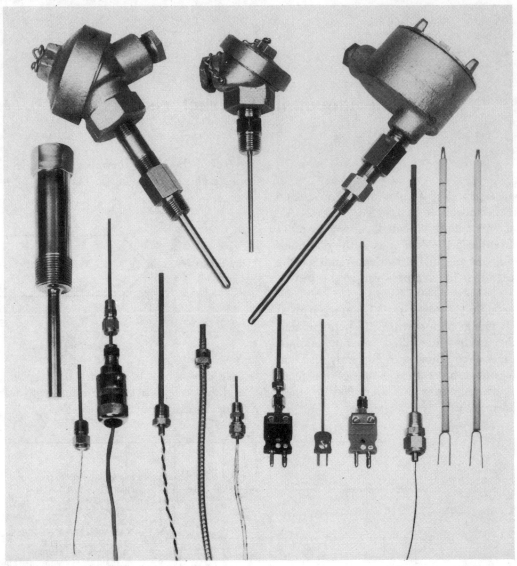

Courtesy Hy-Cal Engineering

Industrial thermocouples

The temperature-millivolt graph in Fig. 2-3B represents typical millivolt outputs for several common themocouples at a wide range of temperatures. On this graph the vertical axis is in millivolts while the horizontal axis is in degrees Fahrenheit. The millivolt output of any thermocouple represented on this graph is listed relative to a reference junction temperature of 32°F. When the reference junction temperature deviates from 32°F, a temperature measurement error exists.

Objective

In this activity you will observe the millivolt output of a thermocouple at different temperatures.

Table 2-3A. Thermocouple Metals and ISA Codes

ISA Code	Metals	
	Positive (+)	Negative (−)
J	Iron	Constantan
K	Chromel	Alumel
E	Chromel	Constantan
T	Copper	Constantan

Equipment

Type J thermocouple
0–100-mV meter
660-W heat cone
120-Vac power supply
Soldering gun
Connecting wires

Procedure

1. Construct the circuit shown in Fig. 2-3C.
2. Connect the 660-W heat cone to 120 Vac and allow it to "warm up" for about 3 minutes. (NOTE: This cone is *very hot*. Extreme caution should be exercised while working with this device.)

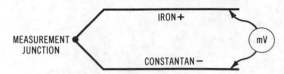

Fig. 2-3C. Operation of type J thermocouple.

3. Place the measurement junction of the thermocouple within ¼ inch of the center of the outside of the cone. Complete the following table by using the temperature-millivolt graph, supplied as a part of this activity, and the temperature conversion formulas from Activity 2-1. (NOTE: The millivolt

Millivolts	°F	°C	K	°R

outputs of the thermocouples listed on the temperature-millivolt graph have reference junction temperatures of 32°F (0°C). Since the temperature of the reference junction of the type J thermocouple used here will be approximately 75°F (23.8°C) a slight measurement error will exist. To convert the type J thermocouple output to reflect a 75°F reference junction, subtract 1.220 mV as indicated in Table 2-3B.

Table 2-3B. Values to Add for Conversion to 75°F Reference Junction

ISA Calibration Symbol	Millivolt Value	ISA Calibration Symbol	Millivolt Value	ISA Calibration Symbol	Millivolt Value
T	−0.947	K	−0.955	R	−0.134
E	−1.427	W	−0.065	S	−0.136
J	−1.220	WS	−0.323	B	0.003

Courtesy CGS/Thermodynamics

4. Disconnect the heat cone and allow it to cool.
5. Connect the soldering gun to 120 Vac, and allow it to "warm up" for about 5 minutes. (NOTE: Extreme caution should be observed while using this device.)
6. Position the measurement junction of the thermocouple until it touches the tip of the soldering gun and complete the following table using the procedure described in Step 3.

Millivolts	°F	°C	K	°R

7. How does the data collected in Step 3 compare with the data collected in Step 6?

_____

_____

_____

Analysis

1. What is a thermocouple?

_____

_____

2. How can a thermocouple be used as a temperature measuring device?

_____

_____

_____

3. What is the relationship between the millivolt output at the reference junction of a thermocouple and the temperature of its measurement junction?

_____

_____

_____

4. What metals are used to form a type T thermocouple?

_____

_____

_____

5. What is the "measurement junction" of a thermocouple?

_____

_____

_____

6. What is the "reference junction" of a thermocouple?

_____

_____

_____

Thermocouple Response Time and
Series and Parallel Connections

Introduction

Thermocouples possess many interesting characteristics that are worthwhile to note. One important characteristic of the thermocouple is its response time. The response time associated with thermocouple action is known as its *time constant* and is defined as the time required for its temperature to reach 63.2 percent of the total "step" change in temperature.

Basically, if the measurement junction of a thermocouple were taken from an environment of 500°F (260°C) and placed in an environment of 1000°F (537.7°C) the "step" change would equal 500°F (1000°F − 500°F = 500°F or 277.7°C difference). If the thermocouple had a time constant of 1 second, then in five time constants or five seconds, its temperature would equal 1000°F. After one second its temperature would equal 816°F or 435.5°C (0.632 × 500°F = 316°F + 500°F = 816°F). Each thermocouple that is manufactured has an assigned time constant that is determined by such characteristics as wire diameter, protective shield, and junction.

Thermocouples may be used singularly or connected in series or parallel as groups called *thermopiles*. When thermocouples are connected in series the resulting output represents the sum of the millivolt outputs of each. When thermocouples are connected in parallel, the output is the mean or average of all of the parallel measurement junctions.

Objective

In this activity you will connect thermocouples in series and in parallel and you will determine the time constant of a thermocouple.

Equipment

Type J thermocouples (2)
0- to 1000-mV meter
600-W heat cone

120-Vac power supply
Connecting wires

Procedure

1. Construct the circuit in Fig. 2-4A.
2. Connect the heat cone to 120 Vac and allow it to "warm up" for about 5 minutes.

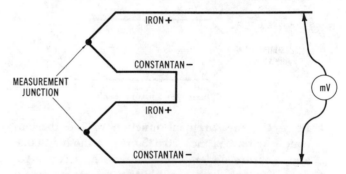

Fig. 2-4A. Two type J thermocouples in series.

3. Place the measurement junctions of the thermocouples illustrated in Step 1 within ¼ inch (0.635 cm) of the heat cone and complete the following table. (NOTE: The temperature-millivolt graph and conversion formulas must be used.)

Millivolts	°F	°C	K	°R

4. Allow the thermocouples to cool and construct the circuit shown in Fig. 2-4B.
5. Position the measurement junctions of the thermocouples shown in Fig. 2-4B within ¼ inch (0.635 cm) of the heat cone and complete the following table employing the same procedure used in Step 3.

Millivolts	°F	°C	K	°R

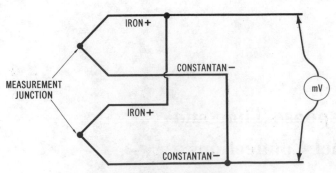

Fig. 2-4B. Two type J thermocouples in parallel.

6. How did the data gathered in Step 3 compare with the data from Step 5?

_____

_____

_____

7. Allow the thermocouples to cool to room temperature.
8. Construct the circuit illustrated in Fig. 2-4C.

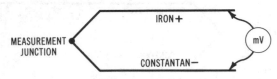

Fig. 2-4C. Basic type J thermocouple.

9. Place the measurement junction of the thermocouple within ¼ inch (0.635 cm) of the heat cone and measure and record the time in seconds required for the thermocouple to reach its maximum temperature.

   $t =$ _____ seconds

10. Using the data gathered in Step 9, determine the time constant exhibited by this thermocouple.

    Time constant = _____ seconds

Analysis

1. What is the relationship that exists between series thermocouples and millivolt output?

_____

2. What relationship exists between parallel thermocouples and millivolt output?

_____

_____

3. What is a thermopile?

_____

_____

4. Where might the series connection of thermocouples be used in an industrial application?

_____

_____

5. Where might the parallel connection of thermocouples be used in an industrial setting?

_____

_____

6. Using the temperature-millivolt graph, determine the millivolt output of four type E thermocouples connected in series whose temperatures are 900 K, 1800°R, 500°C, and 1300°F, respectively.

_____

_____

_____

_____

7. Using the data from the previous question, determine the millivolt output of the above thermocouples connected in parallel.

_____

Thermocouple Applications

Introduction

Thermocouples are frequently used to measure temperatures in an industrial setting. Due to the relatively low voltage output associated with most thermocouples, amplification circuits are used to increase this output as well as to increase sensitivity. The resulting output of the amplification circuit is used to drive or activate a readout device.

Objective

In this activity you will see the thermocouple used to control the conductivity of an FET. The conductivity of the FET, in turn, controls the action of a single-stage transistor amplifier and thus the current flow through a milliampere meter that is used as the readout device.

Equipment

Type J thermocouple
Digital vom
Resistors: 560 Ω, 5.6 kΩ
0–5-Vdc power supply
10-Vdc power supply
GE-FET-1 field-effect transistor
2N2405 npn transistor
Connecting wires
660-W heat cone
120-Vac power supply

Procedure

1. Construct the circuit shown in Fig. 2-5A.
2. Allow the measurement junction of the thermocouple to remain at room temperature. Alter the gate voltage of the FET to equal those listed in Table 2-5A. Record the source-drain current and voltage for each FET gate voltage value.
3. Adjust the gate voltage to 0.2 V.
4. Grasp the measurement junction of the thermocouple between your thumb and forefinger. Describe how this action affects the source-drain

Table 2-5A. Thermocouple-Controlled FET Circuit Data

Gate Voltage (V)	Source-Drain Current (mA)	Source-Drain Voltage (V)
0.2		
0.4		
0.6		
0.8		
1.0		
1.5		
2.0		

current and voltage of the FET as compared to the data gathered in Step 2 when the gate voltage was 0.2 V.

_____

_____

_____

5. Connect the 600-W heat cone to 120 Vac and allow it to "warm up" for about 3 minutes.

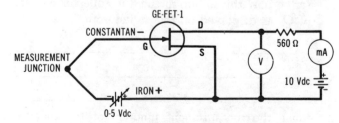

Fig. 2-5A. Type J thermocouple FET circuit.

6. Position the measurement junction of the thermocouple inside the heat cone for a period of 3 minutes and record the source-drain current and voltage of the FET.

$I_{SD} = $ _____; $V_{SD} = $ _____

7. How do the values of source-drain current and voltage recorded in Step 6 compare with the data

gathered in Step 2 when the FET gate voltage was 0.2 Vdc?

8. Disconnect the heat cone from the 120-Vac power supply. What effect does the removal of the heat cone have upon the source-drain current and voltage?

9. Construct the circuit illustrated in Fig. 2-5B.

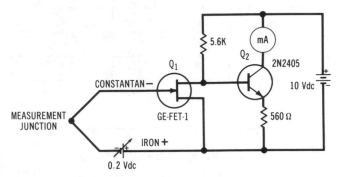

Fig. 2-5B. Type J thermocouple transistor circuit.

10. Record the collector current of Q_2 as displayed by the digital vom with the measurement junction of the thermocouple at room temperature.

$I_C = $ _____

11. Grasp the measurement junction of the thermocouple between your thumb and forefinger. Describe how this action affects the collector current of Q_2 as displayed on the digital vom.

12. Place the measurement junction of the thermocouple inside the cone of the *cool* 660-W heater. Connect the 120 Vac to the heat cone and allow it to "warm up" for 5 minutes.

13. Record the collector current of Q_2 as displayed on the digital vom after the heat cone has "warmed up."

$I_C = $ _____

14. How does the current recorded in Step 10 compare with the current recorded in Step 13?

15. How do you explain the difference?

16. If the digital vom used in the circuit shown in Step 9 were calibrated in degrees Fahrenheit, how could the circuit be used to measure temperature?

Analysis

1. Why is it sometimes necessary to use an amplifier when a thermocouple is used to measure temperature?

2. What was the "readout" device used in the circuit in Step 9?

3. What would determine the maximum temperature that could be measured by the circuit in Step 9?

4. How did the action of the circuit used in Step 9 differ from that used in Step 2?

5. Explain how the circuit in Step 2 could be used to measure temperature.

6. The "readouts" in Steps 10 and 13 were due to current. How could a readout due to voltage change be acquired?

_____

_____

_____

7. Where might a circuit such as the one illustrated in Fig. 2-5B be used in an industrial setting?

_____

_____

_____

Thermistor Characteristics

Introduction

Thermistors are temperature-sensitive resistors that exhibit a negative temperature coefficient. In other words, the electrical resistance of a thermistor will be reduced when it is placed in an environment of higher temperature. Likewise, its resistance is increased when its environmental temperature is decreased. Due to its reaction to temperature and temperature change, the thermistor is widely used in industrial settings as a temperature measuring device.

(a) Directly heated.	(b) Indirectly heated.
HEAT	HEATER

Fig. 2-6A. Thermistor symbols.

Thermistors are manufactured and formed into rods, discs, washers, and beads for special applications. They may be directly or indirectly heated (see Fig. 2-6A). The resistance of those that are directly heated is determined by the temperature of their environment. The resistance of indirectly heated thermistors is determined by the temperature of a self-contained heater.

Three major characteristics of thermistors make them useful electrical devices. These are a thermistor's resistance-temperature, voltage-current, and current-time characteristics.

Objective

You will consider a thermistor's resistance-temperature characteristic and use directly heated devices while doing so.

Equipment

GE-X15 thermistor
Resistor: 1 kΩ
Variable dc power supply
Digital vom

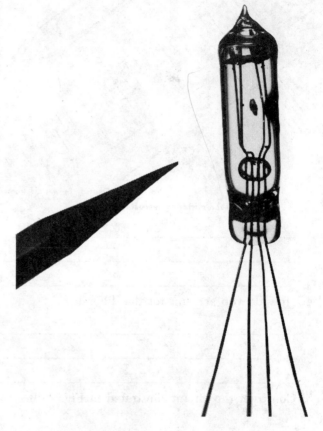

Courtesy Fenwal Electronics

Indirectly heated thermistor.

Connecting wires
Styrofoam cups (2)
Hot and cold water

Procedure

1. Measure and record the resistance of GE-X15 thermistor at room temperature.

   $R =$ _____ Ω

2. Grasp the thermistor with your hand for a period of 3 minutes. Record its resistance after 3 minutes.

   $R =$ _____ Ω

3. How does the resistance recorded in Step 1 compare with the resistance in Step 2?

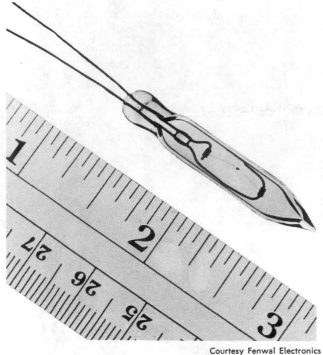

Courtesy Fenwal Electronics

Bead thermistor in evacuated glass bulb.

_____

_____

_____

4. How do you account for the difference?

_____

_____

_____

5. Construct the circuit illustrated in Fig. 2-6B.

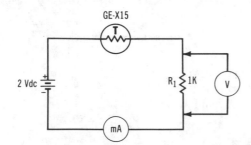

GE-X15

2 Vdc R_1 1K   V

mA

Fig. 2-6B. Thermistor circuit.

6. Record the circuit current and voltage drop across R_1 with the thermistor at room temperature.

   $I =$ _____ mA; $E_{R1} =$ _____ V

7. Submerge the tip of the thermistor into a cup of cold water for a period of 2 minutes and record

the new values of I and E_{R1}.

   $I =$ _____ mA; $E_{R1} =$ _____ V

8. Remove the thermistor from the cold water and allow it to return to room temperature.

9. Submerge the tip of the thermistor into a cup of hot water for a period of 2 minutes and record the circuit current and voltage across R_1.

   $I =$ _____ mA; $E_{R1} =$ _____ V

10. How do the currents and voltages from Steps 6, 7, and 9 compare?

_____

_____

_____

11. How do you account for their being different?

_____

_____

_____

_____

12. Using the data gathered in Steps 6, 7, and 9, complete the following table by computing the resistance of thermistor at room temperature and in hot and cold water.

Environment	Resistance of Thermistor
Room Temperature	
Hot Water	
Cold Water	

13. Using the data computed in Step 12, what is the relation between the temperature of a thermistor and its resistance?

_____

_____

_____

14. What is the relationship between thermistor resistance, thermistor current, and thermistor voltage?

_____

_____

_____

Analysis

1. What is a negative temperature coefficient?

   _____

   _____

   _____

2. What are the common shapes of most commercially available thermistors?

   _____

   _____

   _____

3. What is the difference between a directly and indirectly heated thermistor?

   _____

   _____

   _____

4. What is the relation between thermistor resistance and temperature?

   _____

   _____

   _____

Thermistor Self-Heating

Introduction

The resistance of a thermistor depends upon its temperature. The temperature of any conductor depends upon the amount of current through that conductor. Normally, the greater the current flowing through a conductor, the greater the temperature of the conductor. Thus, as current flows through a thermistor a certain amount of power is converted to heat. This heat tends to lower the resistance of the thermistor, thus allowing current to increase and more power to be converted to heat. This characteristic of the thermistor is known as its *self-heating* characteristic and is usually stated as a dissipation constant.

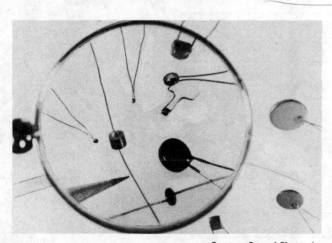

Courtesy Fenwal Electronics

Disc thermistors.

The dissipation constant is important to consider when extreme accuracy in measurement is essential. Basically the dissipation constant of any thermistor is the amount of power in milliwatts that will raise the temperature of the thermistor 1°C above its surroundings.

Objective

In this activity you will observe and graph the self-heating characteristic of a thermistor.

Equipment

GE-X15 thermistor
Resistor: 1 kΩ
Dc power supply
Digital vom
Connecting wires
Spst switch

Procedure

1. Construct the circuit in Fig. 2-7A. Do not close S_1 until you are instructed to do so.

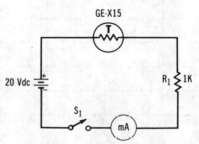

Fig. 2-7A. Thermistor test circuit.

2. Measure and record the resistance of the circuit in Fig. 2-7A.

   $R =$ _____ Ω

3. Using the measured circuit resistance and the indicated power supply voltage, compute the circuit current.

   $I =$ _____ mA

4. Close S_1 and measure the circuit current *after* 5 minutes.

   $I =$ _____ mA

5. How does the current computed in Step 3 compare with the current measured in Step 4?

   _____

   _____

   _____

6. How do you explain the difference in currents in Steps 3 and 4?

_____

_____

7. Open S_1 and allow the thermistor to return to normal.
8. Close S_1 and complete Table 2-7A by recording the current through the thermistor at 20-second intervals for 5 minutes.
9. Using the data from the table in Step 8, complete the graphs in Figs. 2-7B and 2-7C by showing current and resistance versus time.

10. How do the graphs in Step 9 compare?

_____

_____

_____

Analysis

1. Explain the self-heating characteristics of thermistors.

_____

_____

_____

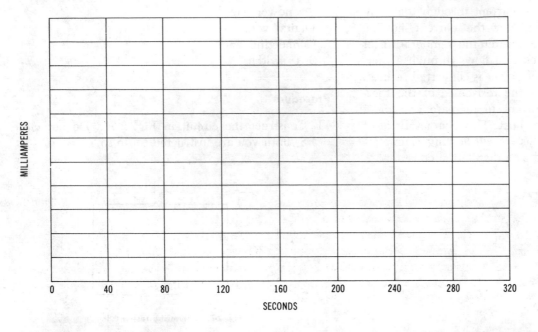

Fig. 2-7B. Thermistor current vs. time.

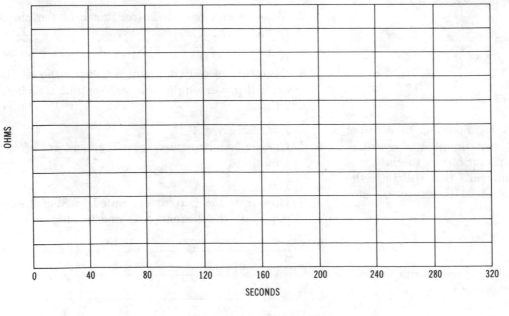

Fig. 2-7C. Thermistor resistance vs. time.

Table 2-7A. Thermistor Circuit Data

Time in Seconds	Thermistor Current Measured (mA)	Thermistor Resistance Computed (Ω)
20		
40		
60		
80		
100		
120		
140		
160		
180		
200		
220		
240		
260		
280		
300		

2. What is the dissipation constant of a thermistor?

3. How is the dissipation constant and thermistor measurement error related?

4. How can self-heating be avoided?

Thermistor Response Time

Introduction

The response time of a thermistor is the amount of time required for a thermistor to heat up and change its resistance. The response time of a thermistor is expressed as a time constant. The time constant of a thermistor is the time required, in seconds, for it to change its temperature (resistance) 63 percent of the value of the temperature change of its environment. Thus, if a thermistor is taken from an environment of 72°F (22.2°C) and placed in an environment of 172°F (77.7°C), the time required for the thermistor to change to have a temperature of 135°F (57.2°C) will be its time constant. (This is easy to see because 172°F − 72°F = 100°F and 63% of 100°F is 63°F. Thus 63°F + 72°F = 135°F.) Normally, five time constants are required for a thermistor to completely reflect a new environmental temperature.

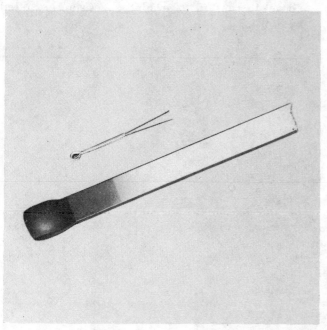

Courtesy Fenwal Electronics

Fast-response micromini thermistor.

Objective

In this activity you will determine the time constant of a thermistor and the change in resistance corresponding to a change in thermistor temperature.

Equipment

GE-X15 thermistor
Fahrenheit thermometer (0°F to 250°F or −17.7°C to 121.1°C)
Resistor: 1 kΩ
Dc power supply
Styrofoam cup
Hot water
Digital vom
Connecting wires
Spst switch

Procedure

1. Place the Fahrenheit thermometer beside the thermistor on the work surface for a period of 5 minutes. Record the (room) temperature of the thermistor as reflected by the thermometer.

   Thermometer temperature = _____ °F
2. Construct the circuit illustrated in Fig. 2-8A. (NOTE: Do not close S_1 until instructed to do so.)
3. Close S_1 and record the circuit current when with the thermistor at room temperature. Open S_1 after recording the circuit current.

   $I =$ _____ mA
4. Acquire a cup of hot water, insert the thermometer into the water and record its temperature after 3 minutes.
   Water temperature = _____ °F
5. Acquire a watch and prepare to measure the time of the next step.
6. Insert the tip of the thermistor into the hot water and record the time required for the thermistor

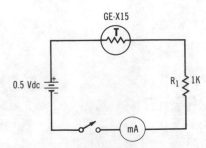

Fig. 2-8A. Circuit for measuring thermistor response time.

to reach its new resistance and thus reflect the temperature of the hot water.

Time = _____ seconds

7. Record the circuit current caused by the resistance of the thermistor in hot water.

$I =$ _____ mA

8. Open S_1, remove both the thermometer and thermistor from the water and allow them to return to room temperature.

9. Using the data collected in Step 6, compute the time constant of the GE-X15 thermistor.

Time constant = _____ seconds

10. Using the data gathered in Step 3 and Step 7, compute the resistance of the thermistor at room temperature and in hot water.

R at room temperature = _____ Ω

R in hot water = _____ Ω

11. Keeping in mind the temperature of the thermistor before and after it was placed in hot water, the circuit current, and the time constant, what was the temperature and resistance of the thermistor after two time constants?

Temperature of the thermistor = _____ °F

Resistance of the thermistor = _____ Ω

Analysis

1. Why is the response time of a thermistor considered important?

_____

_____

2. What is the time constant of a thermistor?

_____

_____

3. How many time constants are required for a thermistor to complete a change in temperature-resistance?

_____

4. The time constant of a thermistor is 0.5 seconds and its resistance is 5000 Ω. If it is taken from an environment of 0°C and placed in an environment of 100°C, what is its temperature and resistance after 1.5 seconds?

_____

5. How many seconds would be required for the above thermistor to exhibit a resistance caused by 100°C?

_____

Thermistor Resistance-Temperature Relationship

Introduction

The resistance of a thermistor is controlled by the temperature of its environment. Thus, if the room-temperature resistance and its temperature are known, its resistance can be determined. Likewise, if room-temperature resistance and its actual resistance are known, its temperature can be determined. These characteristics allow the thermistor, along with the appropriate electrical circuitry, to be used to measure temperature.

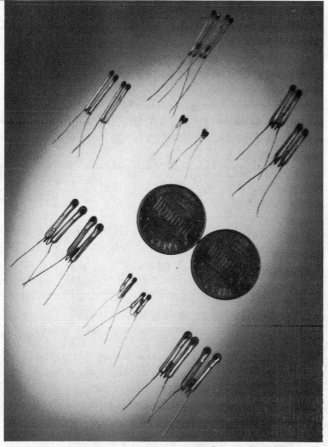

Courtesy Fenwal Electronics

Interchangeable matched thermistors.

A resistance-temperature table like the one illustrated on the following pages can be used to determine the resistance of a thermistor when its temperature ranges from −60°C to 300°C. You will notice that there are 16 different columns of coefficients. Each column represents a different type of thermistor manufactured of a different material.

To use this table, find the temperature of the environment of the thermistor in the left-hand column. Moving your finger horizontally from this temperature to the proper column of coefficients will cause you to identify the correct coefficient to be used to determine the resistance of the thermistor. Then multiply the correct coefficient by the resistance of the thermistor at 25°C (room temperature). The result is the resistance of the thermistor.

Let us assume that we have a 1-kΩ thermistor whose characteristics require that column 10 of the coefficient table be used. Further, let us assume that it is placed in an environment of 100°C. To determine its resistance we find that a coefficient of 0.0946 occurs in column 10 for 100°C. This causes the resistance of our thermistor to be 94.6 Ω when its temperature is 100°C (1000 Ω × 0.0946 = 94.6 Ω).

Assume that the resistance of the same thermistor is determined to be 216 Ω when placed in an environment of an unknown temperature. To determine its temperature, divide its new resistance by its room-temperature resistance to determine the coefficient (215 Ω ÷ 1000 Ω = 0.216). We find 0.216 in column 10 and determine that it coincides with a temperature of 70°C or 158°F.

Objective

In this activity you will compute the new resistance of a thermistor when its temperature is increased to a specified value. You will also use a thermistor resistance value to compute the temperature of thermistor.

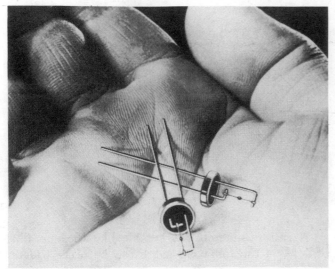

Courtesy Fenwal Electronics

Mounted bead thermistors.

Equipment

Digital vom
Fenwal KA31L1 thermistor (1 kΩ)
Resistor: 1 kΩ
Fahrenheit thermometer (0°F–250°F or −17.7°C to 121.1°C)
Dc power supply
Styrofoam cup
Hot water
Connecting wires
Spst switch
60-W bulb with socket
120-Vac power supply
Box: $4 \times 4 \times 6$ inches ($10 \times 10 \times 15$ cm)

Procedure

1. Construct the circuit in Fig. 2-9A. (Allow S_1 to remain open.)
2. Measure and record the total resistance of the circuit with the thermistor at room temperature.

$R_T =$ _____ Ω

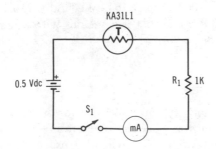

Fig. 2-9A. Thermistor test circuit.

3. Compute the current in the circuit using the measured resistance from Step 2.

Computed $I =$ _____ mA
4. Close S_1, record the circuit current, and open S_1.

Measured $I =$ _____ mA
5. How did the computed current in Step 2 compare with the measured current in Step 4?

_____

_____

6. Acquire a styrofoam cup of hot water, and insert the Fahrenheit thermometer in it. Add cold water to the hot water until its temperature is 122°F (50°C) as measured by the thermometer.
7. Insert the thermistor into the water, close S_1, and record the circuit current. Open S_1.

Measured $I =$ _____ mA
8. Using column 16 of the resistance-temperature table included as a part of this activity, compute the resistance of the thermistor when its temperature is 122°F (50°C).

$R =$ _____ Ω
9. Using Ohm's law and the resistance computed in Step 8, compute the circuit current when the temperature of the thermistor is 122°F (50°C).

Computed $I =$ _____ mA
10. How does the current computed in Step 9 compare with the current measured in Step 7?

_____

_____

11. Connect the 60-W lamp to 120 Vac and place it in a closed $4 \times 4 \times 6$-inch ($10 \times 10 \times 15$ cm) box for approximately 3 minutes.
12. Insert the thermistor through a small hole near the top of the box, close S_1, and record the circuit current after approximately 1 minute.

$I =$ _____ mA
13. Open S_1, remove the thermistor from the small hole, remove the box from the lamp, and disconnect the lamp from its voltage source.
14. Using the current recorded in Step 12, compute the total resistance of the circuit.

Computed $R_T =$ _____ Ω

Table 2-9A. Resistance-Temperature Conversion Table

Table shows curves of thermistors made of different types of materials. To determine resistance of thermistor at specified temperature, first determine RT curve number, material, type unit, and then select appropriate vertical column. Multiply resistance of thermistor at 25 C by appropriate horizontal value in line with the specified temperature to obtain resistance.

R-T CURVE NO.	1	2	3	4	5	6	7
MATERIAL	TYPE T	TYPE P	TYPE H	TYPE H	TYPE H	TYPE H	TYPE H
TYPE UNITS	DISCS	DISCS	STD. LG. BEADS MINI-PROBES STD. PROBES	STD. LG. BEADS MINI-PROBES STD. PROBES	STD. LG. BEADS MINI-PROBES STD. PROBES	STD. LG. BEADS MINI PROBES STD. PROBES	STD. LG. BEADS MINI PROBES STD. PROBES

Ro RANGES (OHMS)

NOTE—FOR DISCS:

Size	DIA. (IN.)
F	.050
J	.1
K	.2
C	.3
L	.4
D	.5
M	.6
N	.77
Z	1.0
*P	.070

NOTE—FOR RODS:

Size	DIA. (IN.)
Q	.053
R	.110
T	.173

Curve 1 DISCS		Curve 2 DISCS		3	4	5	6	7
Size	Ro	Size	Ro	Ro NOM. 300K RANGE 100K-500K	Ro NOM. 500K RANGE 300K-1 MEG.	Ro NOM. 1 MEG. RANGE 600K-3 MEG.	Ro NOM. 5 MEG. RANGE 2 MEG.-10 MEG.	Ro NOM. 50 MEG. RANGE 20 MEG.-80 MEG.
F	45K-180K	F	50K-200K					
J	22K-100K	J	25K-110K					
K	5.5K-50K	K	6K-55K	BEADS .043 DIA.	BEADS .043 DIA.	BEADS .043 DIA.	BEADS .043 DIA.	BEADS .043 DIA.
C	3.5K-24K	C	4K-27K	MINI .060 DIA.	MINI .060 DIA.	MINI .060 DIA.	MINI .060 DIA.	MINI .060 DIA.
L	2K-14K	L	2.2K-15K	STD. .100 DIA.	STD. .100 DIA.	STD. .100 DIA.	STD. .100 DIA.	STD. .100 DIA.
D	1300-9K	D	1400-10K					
M	1100-6K	M	1200-7K					
N	725-3700	N	800-4K					
Z	550-2200	Z	600-2400					
		P	200K-1 MEG.					

PART NUMBERS PREFIXED BY:	FT, JT, KT, CT, LT, DT, MT, NT, UT, ZT	FP, JP, KP, CP, LP, DP, MP, NP, PP, ZP	GH	GH	GH	GH	GH

	1	2	3	4	5	6	7
BETA IN °K	4138±86	4290±100	4227±86	4349±87	4540±86	4850±86	5584±86
RATIO Ro @ 0/50°C	10.45±5%	11.60±4.5%	10.99±5%	11.78±5%	13.12±5%	15.65±5%	23.71±5%
RATIO TEST LIMITS 0/50°C	9.93-10.97	11.08-12.12	10.44-11.54	11.19-12.37	12.46-13.78	14.87-16.43	22.52-24.90
RATIO Ro @ 25/125°C	38.07	48.08	42.20	46.57	56.60	75.50	147.5
TEMPERATURE CO-EFFICIENT (α_T)@25°C	4.7%/°C	4.9%/°C	—4.8%/°C	—4.9%/°C	—5.1%/°C	—5.5%/°C	—6.3%/°C

°F	°C	1	2	3	4	5	6	7
−76	−60	—	—	183.3	201.4	223.9	349.6	455.5
−58	−50		92.08	86.03	94.18	107.4	151.1	205.5
−40	−40	40.155	45.50	42.24	45.95	52.87	68.47	94.97
−22	−30	20.640	23.31	21.61	23.31	26.69	32.41	44.89
− 4	−20	11.034	12.08	11.47	12.22	13.80	15.97	21.68
14	−10	6.119	6.70	6.314	6.642	7.247	8.169	10.69
32	0	3.510	3.71	3.591	3.733	3.942	4.323	5.376
50	10	2.078	2.20	2.107	2.157	2.227	2.354	2.710
68	20	1.2674	1.30	1.272	1.284	1.297	1.322	1.405
77	25	1.0000	1.00	1.000	1.000	1.000	1.000	1.000
86	30	.79422	.796	.7895	.7860	.7764	.7644	.7469
104	40	.51048	.505	.5021	.4934	.4772	.4538	.4068
122	50	.33591	.320	.3267	.3170	.3004	.2762	.2267
140	60	.22590	.212	.2173	.2092	.1936	.1725	.1315
158	70	.15502	.140	.1475	.1409	.1275	.1106	.07831
176	80	.10837	.0957	.1020	.09663	.08562	.07191	.04780
194	90	.077077	.0671	.07178	.06744	.05858	.04786	.02985
212	100	.055693	.0470	.05132	.04784	.04077	.03244	.01904
230	110	.040829	.0337	.03725	.03432	.02882	.02238	.01240
248	120	.030333	.0242	.02743	.02499	.02070	.01569	.008239
257	125	.026266	.0208	.02366	.02144	.01764	.01322	.006764
266	130	.022810	.0178	.02047	.01845	.01508	.01118	.005578
284	140	.017343	.0134	.01546	.01379	.01140	.008077	.003842
302	150	.013319	.0101	.01182	.01044	.008335	.005916	.002690
320	160		*.00768	.009136	.008003	.006325	.004390	.001916
356	180		*.00464	.005629	.004860	.003761	.002506	.001014
392	200		*.00292	.003600	.003071	.002366	.001494	.0005658
428	220		—	.002377	.002012	.001484	.0009256	.0003291
464	240		—	.001619	.001359	.0009767	.0005942	.0002050
500	260		—	.001134	.0009417	.0006622	.0003942	.0001276
536	280		—	.0008156	.0006677	.0004615	.0002694	.0000846
572	300		—	.0006000	.0004836	.0003294	.0001890	.0000581

* 160°C THOUGH 300°C USED FOR PART NUMBERS PREFIXED BY PT, PB & PA ONLY.   **Ro = RESISTANCE @ 25°C, Zero Power Applied

P DIA. REFERS TO GLASS ENVELOPE DIA.

Courtesy Fenwal Electronics

Table 2-9A cont. Resistance-Temperature Conversion Table

Table shows curves of thermistors made of different types of materials. To determine resistance of thermistor at specified temperature, first determine RT curve number, material, type unit, and then select appropriate vertical column. Multiply resistance of thermistor at 25°C by appropriate horizontal value in line with the specified temperature to obtain resistance.

8	9	10	11	12	13	14	15	16
TYPE D	TYPE C	TYPE B	TYPE B	TYPE B	TYPE A	TYPE A	TYPE A	TYPE A
GLASS COATED BEADS & PROBES DISCS	GLASS COATED BEADS & PROBES	DISCS WASHERS RODS	GLASS COATED BEADS & PROBES	GLASS COATED BEADS & PROBES	GLASS COATED BEADS & PROBES	GLASS COATED BEADS & PROBES	GLASS COATED BEADS & PROBES	DISCS WASHERS RODS
STD. SMALL BEADS (.014 DIA.) 250—1K; STD. LG. BEADS & PROBES (.043 DIA.) 10-250; DISCS Size Ro F 30-50 J 15-75 K 4-35 C 2.5-18 L 1.5-10 D .9-6.5 M .7-4.5; PROBES MICRO-MINI (.020 DIA.) 250—1K; SUB-MINI (.030 DIA.) 150-650; MINI (.060 DIA) 50-250; GD BEADS & PROBES -- DISCS -- FD, JD, KD, CD, LD, DD, MD, UD	STD. SMALL BEADS (.014 DIA.) 1K-5K; STD. LG. BEADS & PROBES (.043 DIA.) 250-2K; PROBES MICRO-MINI (.020 DIA.) 1K-5K; SUB-MINI (.030 DIA.) 600-2K; GC	DISCS Size Ro F 600-2800 J 300-1400 K 75-700 C 50-350 L 30-180 D 20-125 M 15-85 N 10-50 Z 7.5-30 P 2.5K-15K; WASHERS 10-60; RODS Size Ro Q 4K-20K R 1K-15K T 350-7.5K; DISCS FB, JB, KB, CB, LB, DB, MB, NB, PB, UB, ZB; WASHERS WB; RODS GB, RB, TB	STD. SMALL BEADS (.014 DIA) 7K-30K; STD. LG. BEADS (.043 DIA.) 1K-5K; PROBES MICRO-MINI (.020 DIA.) 7K-30K; SUB-MINI (.030 DIA.) 4K-18K; MINI (.060 DIA.) 1K-5K; STD. PROBES (.100 DIA.) 1K - 5K; GB	STD. SMALL BEADS (.014 DIA) 40K-50K; STD. LG. BEADS (.043 DIA.) 5K-10K; PROBES MICRO-MINI (.020 DIA.) 40K-50K; SUB-MINI (.030 DIA.) 23K-30K; MINI (.060 DIA.) 5K-10K; STD. PROBES (.100 DIA.) 5K - 10K; GB	STD. SMALL BEADS (.014 DIA.) 50K-200K; STD. LG. BEADS (.043 DIA.) 10K-30K; PROBES MICRO-MINI (.020 DIA.) 50K-200K; SUB-MINI (.030 DIA.) 30K-120K; MINI (.060 DIA.) 10K-30K; STD. PROBES (.100 DIA.) 10K - 30K; GA	STD. SMALL BEADS (.014 DIA.) 200K-400K; STD. LG. BEADS (.043 DIA.) 30K-60K; PROBES MICRO-MINI (.020 DIA.) 200K-400K; SUB-MINI (.030 DIA.) 110K-230K; MINI (.060 DIA.) 30K-60K; STD. PROBES (.100 DIA.) 30K - 60K; GA	STD. SMALL BEADS (.014 DIA.) 500K-1 MEG.; STD. LG. BEADS (.043 DIA.) 75K-200K; PROBES MICRO-MINI (.020 DIA.) 500K-1 MEG.; SUB-MINI (.030 DIA.) 280K-600K; MINI (.060 DIA.) 75K-200K; STD. PROBES (.100 DIA.) 75K - 200K; GA	DISCS Size Ro F 4400-20K J 2200-10K K 550-5K C 375-2500 L 200-1400 D 130-900 M 110-600 N 72-375 Z 55-220 P 20K-100K; WASHERS 70-425; RODS Size Ro Q 25K-125K R 6K-120K T 2.5K-42.5K; DISCS FA, JA, KA, CA, LA, DA, MA, NA, PA, UA, ZA; WASHERS WA; RODS QA, RA. TA
2758±175	3000±175	3400±80	3442±90	3574±93	3894±90	3976±93	4118±95	3887±51
4.80±10%	5.50±10%	6.95±4.5%	7.04±5%	7.59±5%	9.1±5%	9.53±5%	10.45±5%	9.1±3%
4.32-5.28	4.95-6.05	6.63-7.26	6.69-7.39	7.21-7.97	8.65-9.56	9.05-10.01	9.93-10.97	8.83-9.37
10.30	13.51	19.05	19.85	22.73	29.42	31.72	38.05	29.27
−3.1%/°C	−3.4%/°C	−3.9%/°C	−3.9%/°C	−4.0%/°C	−4.4%/°C	−4.5%/°C	−4.7%/°C	−4.4%/°C
38.2	45.3	73.04	76.08	89.45	145.2	152.5	174.0	140.49
21.7	25.8	38.95	40.10	46.03	68.88	72.00	81.6	67.01
12.90	15.1	21.51	22.07	24.75	34.28	37.268	40.2	33.65
8.03	9.24	12.33	12.60	13.83	17.92	18.40	20.6	17.70
5.16	5.81	7.307	7.430	8.009	9.792	10.20	11.0	9.707
3.42	3.76	4.476	4.530	4.796	5.560	5.767	6.12	5.533
2.34	2.50	2.825	2.850	2.961	3.274	3.363	3.51	3.265
1.64	1.70	1.830	1.839	1.882	1.992	2.022	2.08	1.990
1.17	1.19	1.216	1.219	1.227	1.250	1.256	1.27	1.249
1.00	1.00	1.000	1.000	1.000	1.000	1.000	1.00	1.000
.857	.846	.8267	.8265	.8197	.8053	.8030	.794	.8057
.640	.615	.5742	.5730	.5598	.5316	.5264	.510	.5327
.486	.454	.4067	.4048	.3903	.3595	.3528	.336	.3603
.376	.341	.2937	.2915	.2773	.2482	.2417	.226	.2488
.295	.261	.2160	.2138	.2006	.1747	.1690	.155	.1752
.234	.202	.1615	.1594	.1475	.1252	.1203	.108	.1255
.189	.158	.1229	.1205	.1101	.09126	.08698	.0771	.09153
.154	.125	.0946	.09235	.08335	.06754	.06395	.0557	.06783
.127	.101	.0740	.07185	.06396	.05076	.04769	.0408	.05103
.106	.0817	.0585	.05655	.04969	.03867	.03608	.0303	.03893
.0971	.0740	.0525	.05038	.04399	.03399	.03154	.0262	.03417
.0889	.0670	.0471	.04500	.03906	.02988	.02765	.0228	.030093
.0755	.0554	.0382	.03620	.03104	.02327	.02144	.0173	.023527
.0647	.0462	.0314	.02940	.02491	.01843	.01682	.0133	.018597
		*.0259	.02408	.02019	.01470	.01332	.0105	*.0147
		*.0180	.01727	.01362	.009700	.008615	.00656	*.0097
		*.0130	.01248	.009491	.006600	.005769	.00427	*.0066
		—	.00940	.006805	.004700	.003981	.00286	—
		—	.007294	.005004	.003500	.002831	.00197	—
		—	.005758	.003763	.002600	.002065	.00140	—
		—	.004607	.002888	.002100	.001541	.00101	—
		—	.003839	.002256	.001700	.001173	.000745	—

Courtesy Fenwal Electronics

15. Using the data gathered in Steps 2 and 14, compute the new resistance of the thermistor.

    R of thermistor = _____ Ω

16. Using column 16 of the resistance-temperature table included as a part of this activity, and the computed new resistance of the thermistor in Step 15, compute the temperature caused by the 60-W lamp in the box.

    Temperature = _____ °F

    = _____ °C

    = _____ K

    = _____ °R

Analysis

1. How can a thermistor, along with the appropriate electrical circuitry, be used to measure temperature?

    _____

    _____

2. How can a resistance-temperature table be used to determine the resistance of a thermistor at a specified temperature?

    _____

    _____

    _____

3. How can a resistance-temperature table be used to determine the temperature of a thermistor at a specified resistance?

    _____

    _____

    _____

4. Where might the thermistor be used to measure temperature in an industrial setting?

    _____

    _____

Thermistor Applications

Introduction

Thermistors are generally used in conjunction with other electrical components and circuitry for most applications. The most common combination of components employs a thermistor to control the action of an amplifier which, in turn, provides an output to drive a readout device. The readout device indicates current controlled by the resistance-temperature of the thermistor and may be calibrated in kelvins or degrees Fahrenheit, Celsius, or Rankine.

Another very popular configuration using thermistors is the Wheatstone bridge circuit. Here a thermistor may be used in one or more arms of the bridge circuit to measure temperature, thermal conductivity, altitude, and wind velocity.

Objective

In this activity you will examine some common configurations using the thermistor.

Equipment

Digital vom
Thermistors: Fenwal JA41J1 and JA35J1 (2)
Dc power supply
Transistors: 2N2405
Resistors: 680 Ω, 10 kΩ
Potentiometers: 25 kΩ, 2.5 kΩ
Connecting wires
Fahrenheit thermometer (0°F–250°F or −17.7°C– 121.1°C)
Spst switch
Styrofoam cup
Hot water

Procedure

1. Construct the circuit illustrated in Fig. 2-10A.
2. Close S_1 and adjust R_1 until the milliammeter indicates zero current when the thermistor is at room temperature or 72°F (22.2°C). (NOTE: R_1

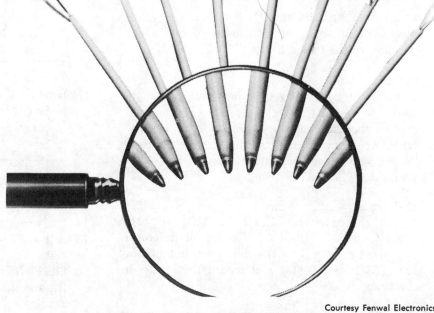

Medical thermistor probes.

Courtesy Fenwal Electronics

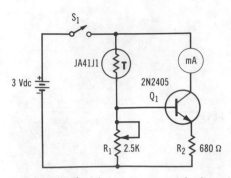

Fig. 2-10A. Thermistor measurement circuit.

should be adjusted until the milliammeter indicates 1 µA of collector current. Then R_1 should be *slowly* adjusted until the milliammeter indicates that Q_1 is, in fact, exhibiting zero collector current.)

3. Grasp the thermistor between your thumb and forefinger for about 3 minutes and record the current by the milliammeter.

$I =$ _____ mA

4. Open S_1 and explain why there is a collector current when the thermistor is grasped between your fingers.

_____

_____

_____

5. Acquire a styrofoam cup and fill it half full of hot water. Insert the thermometer into the hot water and add cool water until the thermometer indicates the water temperature to be 125°F (51.6°C).

6. Insert the thermistor into the water, close S_1, and record the collector current caused when the temperature of the thermistor is 125°F (51.6°C).

$I =$ _____ mA

7. Open S_1 and remove the thermistor from the water.

8. Record the difference between collector currents when the temperature of the thermistor is at 72°F (Step 2) and when the temperature of the thermistor is at 125°F (Step 6).

Difference = _____ mA

9. The difference between 125°F (51.6°C) and 72°F (22.2°C) is 53°F (29.4°C). Divide the difference in current in Step 8 by the difference in temperature (53°F or 29.4°C) and record the results in the space below.

_____

10. The results in Step 9 represents the amount of collector current increase for each degree-Fahrenheit increase in the temperature of the thermistor above 72°F or 22.2°C.

11. Using the result of Step 9, compute the collector current when the temperature of the thermistor is 90°F, 100°F, 110°F, and 120°F (32.2°C, 37.7°C, 43.3°C, 48.8°C).

Thermistor Temperature (°F)	Computed Collector Current
90	
100	
110	
120	

12. Acquire a styrofoam cup of hot water, insert the thermometer, insert the thermistor, and close S_1. Add cool water to the hot water to cause the water temperature to equal 120°F, 110°F, 100°F, and 90°F (48.8°C, 43.3°C, 37.7°C, 32.2°C) as indicated by the thermometer. Record in the following table the collector current caused when the thermistor is at each of these temperatures.

Thermistor Temperature (°F)	Measured Collector Current
90	
100	
110	
120	

13. How do the computed values of collector current in Step 1 compare with the measured values in Step 12?

_____

_____

_____

14. Construct the circuit shown in Fig. 2-10B.

15. Adjust R_3 until the digital vom indicates the "null" state.

16. Shield R_1 while allowing a "breeze" to pass over R_4, and describe how this affects the digital vom's reading.

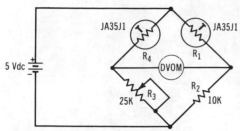

Fig. 2-10B. Thermistor bridge circuit.

_____

_____

17. Shield R_4 while allowing a "breeze" to pass over R_1, and describe how this affects the digital vom's reading.

_____

_____

_____

18. What is the difference between effects on the vom's reading in Steps 16 and 17?

_____

_____

_____

19. What happens when the intensity of the "breeze" in Step 16 or Step 17 is increased?

_____

_____

_____

20. What is an anemometer?

_____

_____

_____

Analysis

1. Why is it necessary to use an amplifier with a thermistor for certain applications?

_____

_____

_____

2. Analyze the operation of the circuit in Step 1 of this activity.

_____

_____

_____

_____

_____

3. Why does the Wheatstone bridge circuit work so well with thermistors?

_____

_____

_____

4. How could a thermistor and the appropriate circuitry be used to measure wind velocity?

_____

_____

_____

5. How could a thermistor, along with an electrical circuit, be used to measure thermal conductivity?

_____

_____

_____

_____

Resistance Thermometer

Introduction

The resistance thermometer is made up of a transducer and an electrical circuit. Its purpose is to measure temperature that causes a change in resistance. The most common electrical arrangement related to the construction of the resistance thermometer allows the temperature transducer to represent one arm of a four-arm resistance bridge circuit.

Objective

In this activity you will examine the thermistor used with a four-arm bridge circuit. This resistance thermometer allows accurate temperature measurement from 0°C to 50°C.

Equipment

Digital vom
Resistors: 1 kΩ (2), 365 Ω

Potentiometers: 1.5 kΩ, 2.5 kΩ, 5 kΩ, 500 kΩ
Decade resistance box
GE-X15 thermistor
Spst switch
Dc power supply
Connecting wires
Styrofoam cup
Crushed ice
Fahrenheit thermometer (0°F–250°F or −17.7°–121.1°C)

Procedure

1. Construct the circuit in Fig. 2-11A, allowing S_1 and S_2 to remain open. (NOTE: The values of the components are critical. Be sure that all selected components have values as close as possible to those shown to ensure accuracy.)
2. The total resistance of the null branch, including the digital vom and R_6 should be as close to 1500 Ω as possible. In order to adjust R_6 to the value

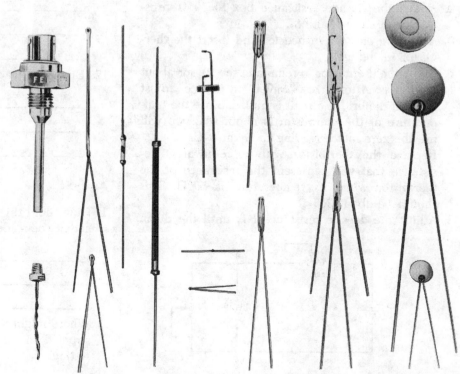

Thermistor probes.

Courtesy Fenwal Electronics

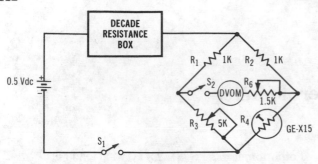

Fig. 2-11A. Resistance thermometer.

that will cause the null branch to equal 1500 Ω of resistance, the resistance of the digital vom must be determined. The following steps should be carefully completed to determine this resistance:
(a) Adjust the vom to the 1-mA range.
(b) Construct the circuit illustrated in Fig. 2-11B.
(c) Adjust R_7 until the dvom indicates 50 μA when on the 1-mA range.
(d) Connect a 1.5-kΩ potentiometer (R_8) to points A and B as illustrated in Fig. 2-11B.
(e) Adjust R_8 until the dvom indicates 25 μA.
(f) Disconnect R_8 from points A and B and measure its resistance.
(g) This resistance is equal to the dvom resistance.

3. Knowing the meter resistance as determined from the previous steps in Procedure Step 1, adjust R_6 until the combination resistance of the meter and R_6 equals 1500 Ω.

4. Adjust the decade resistance box to 50 Ω resistance and close S_1 and S_2.

5. Acquire a cup of crushed ice and insert the thermistor in the ice.

6. Allow the thermistor to remain in the ice for about 3 minutes. After it has "cooled" in the ice, adjust R_3 until the bridge circuit is nulled. Since the temperature of the thermistor is about 0°C, we will use the zero meter reading to equal 0°C.

7. Replace the thermistor with an equivalent resistance that will represent the resistance of the thermistor when its temperature is 50°C. This should be about 365 Ω.

8. Adjust the decade resistance box until the dvom

indicates 50 μA of current. This is the reading the dvom should indicate when the temperature of the thermistor equals 50°C. Replace the 365-Ω resistor with the thermistor.

9. You have now calibrated the resistance thermometer to cause 1 μA of current to equal 1°C.

10. To determine the accuracy of your thermistor, measure the following with a mercury thermometer: the temperature of cold water, warm water, and the human body. Place the thermistor in the same environment and compare the readings. Complete Table 2-11A while engaged in these activities.

Table 2-11A. Resistance Thermometer Data

Environment	Mercury Thermometer Reading (°F)	Null Meter (μA)	Bridge Thermometer Reading			
			°F	°C	K	°R
Hot Water						
Cold Water						
Human Body						

11. How do the readings provided by the resistance thermometer compare with the Fahrenheit thermometer readings?

_____

_____

_____

12. How do you account for any differences?

_____

_____

_____

Analysis

1. How could the resistance thermometer used in this activity be made more accurate?

_____

_____

2. Where might a thermometer, such as the one constructed in this activity, be used in an industrial setting?

_____

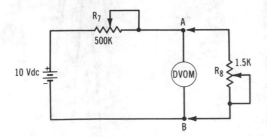

Fig. 2-11B. Circuit to determine dvom resistance.

_____

3. Could a device such as the resistance thermometer be used to measure temperatures in excess of 50°C?

_____

_____

4. Why does the four-arm bridge circuit work well with the thermistor when measuring small temperature variations?

_____

5. Briefly explain how the thermometer examined in this activity could be recalibrated to measure temperatures higher than 50°C.

_____

_____

_____

_____

_____

RTD Characteristics

Introduction

Resistance temperature detectors are thermally sensitive resistive elements that exhibit an increase in resistance as the temperature of their environment increases. Thus, these devices have a positive temperature coefficient, and they are constructed of platinum, nickel, copper, tungsten, or nickel-iron.

Since the rtd will increase its resistance when the temperature of its environment increases, its resistance is a function of temperature and its usually determined at 0°C. Thus a 100-Ω rtd will exhibit 100 Ω of resistance when its temperature is 0°C, or 32°F. As its temperature increases, likewise its resistance increases. To determine the resistance of an rtd at any temperature within its range, the rtd's temperature coefficient must be used. The positive temperature coefficient of resistance is stated as *alpha* and represents the percent of change in resistance for each degree-Celsius change in temperature.

The alpha of a platinum rtd is listed as 0.00385. If a platinum rtd of 100 Ω were removed from an environment of 0°C and placed into a new environment of 100°C, its resistance would increase from 100 Ω to 138.5 Ω. [The 100 Ω of resistance (at 0°C) is multiplied by alpha (0.00385), and this product multiplied by the 100°C difference between the old environment and the new environment, and then added to the 100 Ω resistance at the old environment, which equals 138.5 Ω.]

$$(100\,\Omega \times 0.00385 \times 100) + 100\,\Omega = 138.5\,\Omega$$

Objective

In this activity you will examine the operating characteristics of an rtd.

Equipment

Digital vom
Rtd: platinum, 100 Ω @ 0°C
Resistor: 100 Ω
Ac-dc power supply

Courtesy Hy-Cal Engineering

Resistance temperature sensors (rtd's).

660-W heat cone
Connecting wires

Procedure

1. Construct the circuit illustrated in Fig. 2-12A.
2. Record the current of the circuit with the rtd at room temperature.

$I =$ _____ mA

3. Compute the resistance of the rtd using the current as measured in Step 2.

Resistance of rtd = _____ Ω

Courtesy Hy-Cal Engineering

Resistance temperature elements.

4. Place the 660-W heat cone within 1 inch (2.54 cm) of the sensing end of the rtd. Connect 120 Vac to the heat cone and allow the heat cone to "warm up" for a period of 3 minutes.

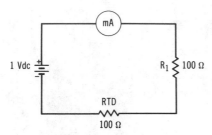

Fig. 2-12A. Rtd series circuit.

5. Record the circuit current after the rtd has "warmed" for 3 minutes.

   $I =$ _____ mA

6. Compute the resistance of the rtd using the current as measured in Step 5.

   Resistance of rtd = _____ Ω

7. How does the current in Step 2 compare with the current measured in Step 5?

   _____

   _____

   _____

8. How does the resistance of the rtd computed in Step 3 compare with its resistance in Step 6?

   _____

   _____

   _____

9. How could the circuit in Step 1 be used to measure an unknown temperature?

   _____

   _____

   _____

Analysis

1. What is the relationship between an rtd's resistance and its temperature?

   _____

   _____

   _____

2. What is a positive temperature coefficient?

   _____

   _____

   _____

3. The resistance of most rtd's is rated at what temperature?

   _____

4. If the alpha of a platinum 100-Ω rtd is 0.00385 and it is placed in an environment whose temperature is 50°C, what is its resistance?

   _____

5. What is the difference between the operating characteristics of a thermistor and an rtd?

   _____

   _____

   _____

   _____

RTD Applications

Introduction

The resistance change in rtd's caused by a change in temperature is very linear. This characteristic allows the rtd to be used, with the appropriate electrical circuitry, to measure the temperature.

Rtd's may be used to control the action of certain types of dc amplifiers when increased sensitivity and greater outputs are required. The most common form of electrical circuitry used in conjunction with the rtd is the four-arm resistive bridge. The most sensitive electrical arrangement is realized when the four-arm bridge is used to control the output of a dc amplifier.

Objective

In this activity you will examine the rtd used with appropriate electrical circuitry to measure temperature.

Equipment

Digital vom
Rtd: platinum, 100 Ω @ 0°C
Resistors: 100 Ω (2), 680 Ω, 6.8 kΩ
Potentiometer: 200 Ω, 1 W
Ac-dc power supplies (2)

Industrial resistance temperature sensors.

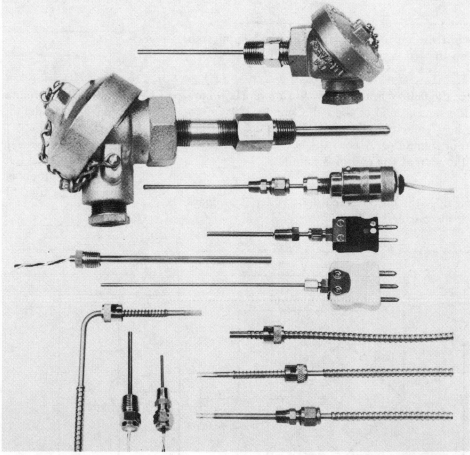

Courtesy Hy-Cal Engineering

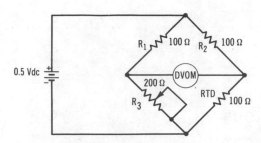

Fig. 2-13A. Rtd bridge circuit.

GE-FET-1 field-effect transistor
Transistor: 2N2405
660-W heat cone
Connecting wires

Procedure

1. Construct the circuit in Fig. 2-13A.
2. Adjust R_3 until the null state is indicated by the digital vom.
3. Grasp the rtd in your hand and describe how this affects the null state.

_____

_____

4. Record the current displayed by the digital vom when the rtd is grasped in your hand.

   $I =$ _____ mA

5. Place the 660-W heat cone within 1 inch (2.54 cm) of the sensing end of the rtd. Connect 120 Vac to the heat cone and allow it to "warm up" for a period of 3 minutes. Record the current as indicated by the vom at the end of 3 minutes.

   $I =$ _____ mA

6. How does the current recorded in Step 3 compare with the current in Step 5?

_____

_____

7. Disconnect the heat cone and allow the rtd to cool in order to cause the bridge circuit to return to the null state.
8. Alter the circuit in Step 1 to the circuit illustrated in Fig. 2-13B.
9. Grasp the rtd in your hand and record the collector current of Q_2 caused by this action.

   $I =$ _____ mA

10. Place the 660-W heat cone within 1 inch (2.54 cm) of the sensing end of the rtd. Connect 120 Vac to the heat cone and allow it to "warm up" for a period of 3 minutes. Record the collector current of Q_2 after 3 minutes.

    $I =$ _____ mA

11. How did the current recorded in Step 4 compare with the current recorded in Step 9?

_____

_____

_____

12. How did the current recorded in Step 5 compare with the current recorded in Step 10?

_____

_____

_____

13. Which of the circuits illustrated in Steps 1 and 8 would be considered more sensitive?

_____

14. If the meters used in Steps 1 and 8 were calibrated in degrees Fahrenheit, how could the circuits in these steps be used to measure temperature?

_____

_____

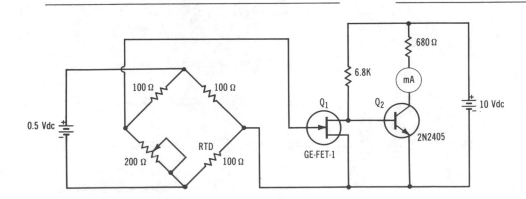

Fig. 2-13B. Rtd bridge and amplifier circuit.

Analysis

1. Write an analysis of the operation of the circuit illustrated in Step 8.

_____

_____

_____

_____

_____

_____

_____

_____

_____

_____

_____

_____

_____

_____

2. Why is it advantageous to use an amplifier with the rtd bridge?

_____

_____

_____

_____

Lead Temperature Compensation

Introduction

Throughout this unit you have examined several types of transducers that are used in the measurement of temperature. It is obvious that in most temperature measurement applications the transducer is placed several feet from its electrical circuitry and readout device. This means that the transducer must be connected to its circuitry by conductors or leads.

The resistance of any lead is affected by the temperature of its environment. Temperatures in an industrial environment vary greatly. Thus the resistance variation of transducer leads caused by temperature variation can result in measurement error. Likewise, when thermocouples are used with connecting leads, each connection between the thermocouple and the lead forms a junction. This "connection junction" can cause a "secondary measurement junction" which generates a millivolt output proportional to the temperature of its environment. This millivolt output may be in series to aid or oppose the initial output of the thermocouple, causing a measurement error.

Most manufacturers of temperature sensing transducers provide special lead connections, lead wire, or lead junctions to be used with their products to offset measurement error. Rtd's are provided with two-, three-, or four-wire lead configurations to compensate for measurement error.

Objective

In this activity you will study the effects of extra-long leads that connect the thermocouple with the output measurement device.

Equipment

Digital vom
Type J thermocouple
660-W heat cone
Ac power supply
Styrofoam cups (2)
Crushed ice
Connecting wires

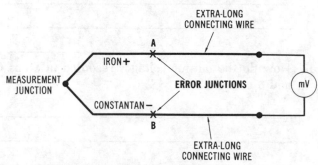

Fig. 2-14A. Type J thermocouple with long wires.

Procedure

1. Construct the circuit in Fig. 2-14A.
2. The extra-long connecting wires represent leads from the environment being measured to the readout device (millivoltmeter).
3. Place the measurement junction of the thermocouple inside the 660-W heat cone. Connect 120 Vac to the heat cone and allow it to "warm up" for 5 minutes.
4. Record the millivolt output of the thermocouple after it has reached its maximum and has stabilized.

   Output = _____ mV
5. Fill the two styrofoam cups with a mixture of ice and water (more ice than water).
6. Immerse error junction A into one of the ice baths and record the output of the thermocouple indicated by the digital vom.

   Output = _____ mV
7. Remove error junction A from the ice bath and allow it to return to room temperature.
8. Immerse error junction B into one of the ice baths and record the output of the thermocouple.

   Output = _____ mV
9. Remove error junction B from the ice bath and allow it to return to room temperature.
10. Immerse both error junctions A and B in separate ice baths and record the output of the thermocouple.

    Output = _____ mV

11. How does the output voltage recorded in Step 4 compare with that recorded in Step 6?

_____

_____

_____

12. How does the output voltage recorded in Step 4 compare with that recorded in Step 8?

_____

_____

_____

13. How do the output voltages recorded in Steps 6 and 8 compare?

_____

_____

_____

14. How do the output voltages recorded in Steps 4 and 10 compare?

_____

_____

_____

15. How does the temperature of the junctions formed by extension wires affect the voltage output of the measurement system?

_____

_____

_____

Analysis

1. What is meant by the term "error junction"?

_____

_____

_____

2. Why must "connection junctions" or "error junctions" be held at constant temperatures?

_____

_____

_____

3. What effects can a voltage generated at an error junction have upon the final output of a measurement system?

_____

_____

_____

4. Why is it important to use carefully selected lead connections and wire when the environment being measured is some distance from the indicator or readout device?

_____

_____

_____

_____

_____

Photoelectric Transducers for Instrumentation Systems

Photoelectric transducers are a very important type of transducer which is used with instrumentation systems and many other applications. These types of transducers have recently received emphasis for many measurement applications. Photoelectric transducers are light-sensitive devices that convert changes of light energy into changes of electrical energy. As an example, when a photovoltaic cell is exposed to light, it will develop a voltage output. Other photoelectric transducers convert changes of light into changes of resistance. These are called *photoresistive* devices.

Courtesy Mikron Instrument Co.

Handheld infrared thermometer.

Photoelectric devices have typically been considered as falling into three categories: (1) photoemissive, (2) photoconductive, and (3) photovoltaic. Photoemissive devices emit electrons in the presence of light. Phototubes are a type of photoemissive device. Photoconductive devices are designed so that their resistance will decrease when light becomes more intense and increase when light intensity decreases. Photoconductive devices are also called *photoresistive*. Photovoltaic devices convert light energy into electrical energy. When a photovoltaic device is illuminated, an electrical potential is created by the device. Most photoelectric devices fit into one of these categories. However, there is such a diversity of new semiconductor photoelectric devices used today that it is more desirable to study each device individually. The activities which follow will deal with various types of photoelectric devices which have applications in instrumentation.

The activities of Unit 3 are developed to provide an understanding of photoelectric transducers and their associated circuitry. Activity 3-1 is used as a general introduction to Unit 3 by discussing character-

Courtesy Mikron Instrument Co.

Infrared thermometer with scope for precision sighting.

istics of light and light sources. Activities 3-2 and 3-3 deal with photoconductive transducers. Photoemissive transducers are studied in Activities 3-4 and 3-5, while Activities 3-6 and 3-7 investigate photovoltaic transducers. Several activities in Unit 3 deal with circuitry used with photoelectric transducers in instrumentation systems. These are primarily pulse modulation circuits. The last activity of Unit 3 provides an understanding of optical couplers, which are becoming increasingly important in electrical instrumentation systems.

Characteristics of Light

Introduction

Light is a visible form of radiation which is actually a narrow band of frequencies along the vast electromagnetic spectrum. The electromagnetic spectrum, shown in Fig. 3-1A, includes bands of frequencies for radio, television, radar, infrared radiation, visible light, ultraviolet light, X-rays, gamma rays, and various other frequencies. The different types of radiation, such as light, heat, radio waves, and X-rays, differ only with respect to their frequencies or wavelengths.

The human eye responds to electromagnetic waves in the visible light band of frequencies. Each color of light has a different frequency or wavelength. In order of increasing frequency, or decreasing wavelengths, colors range as follows: red, orange, yellow, green, blue, and violet. The wavelengths of visible light are in the 400-nanometer (violet) to 700-nanometer (red) range. A micrometer (μm) is one millionth of a meter, and a nanometer (nm) is 1×10^{-3} micrometer. Angstrom units (Å) are also used for light measurement. An angstrom unit is one-tenth of a nanometer. Thus visible light ranges from 4000 Å to 7000 Å. The response of the human eye to visible light exhibits a frequency selective characteristic, such as shown in Fig. 3-1B. The greatest sensitivity is near 5500 Å, and the poorest sensitivity is around 4000 Å on the lower wavelengths and 7000 Å on the higher wavelengths. Our eyes perceive various degrees of brightness due to their response to the wavelengths of light. The normal human eye cannot see a wavelength of less than 4000 Å or more than 7000 Å (400–700 nm).

When dealing with light, there are several characteristic terms which should be understood. The unit of luminous intensity is a standard light source called a *candela*. Thus the intensity of light is expressed in

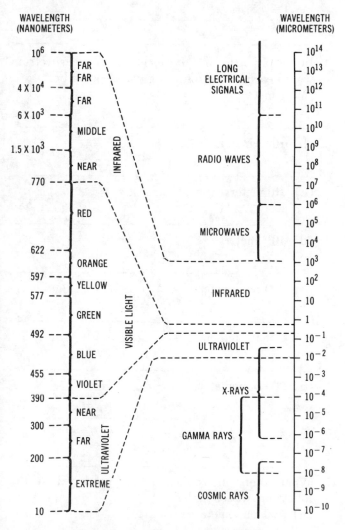

Fig. 3-1A. The electromagnetic spectrum.

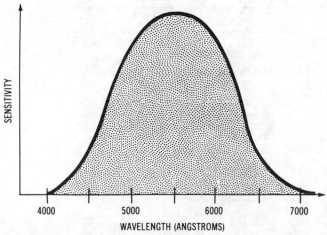

Fig. 3-1B. Response of the human eye to visible light.

candelas. The amount of light falling on a unit surface all points of which are unit distance from a uniform light source of one candela is one *lumen*. The illumination of a surface is the number of lumens falling on it per unit area. The unit of illumination is the *lux* (lumen per square meter).

We see only light which is reflected. Reflected light is measured in candelas per square meter of surface. *Reflection factor* is the percent of light reflected from a surface expressed as a decimal. Therefore the light reflected from a surface is equal to the illumination of the surface times the reflection factor.

The preceding terms are used extensively in discussions of light characteristics and light sources.

Objective

In this activity you will become familiar with the characteristics of light by completing the questions and problems in the *Analysis* section of this activity.

Analysis

1. What is meant by the term *photoelectric?*

   _____

   _____

   _____

2. What is the electromagnetic spectrum?

   _____

   _____

   _____

3. Define the following terms:
   (a) Micrometer:

   _____

   _____

   (b) Millimicrometer:

   _____

   _____

   _____

   (c) Angstrom:

   _____

   _____

   _____

(d) Candela:

   _____

   _____

   _____

(e) Nanometer:

   _____

   _____

   _____

(f) Lumen:

   _____

   _____

   _____

(g) Lux:

   _____

   _____

   _____

4. In what portion of the electromagnetic spectrum would each of the following wavelengths be?
   (a) 1 meter:

   _____

   (b) 10^{-8} meter:

   _____

   (c) 10^5 meters:

   _____

   (d) 10^{12} meters:

   _____

   (e) 500 nanometers:

   _____

   (f) 6×10^3 nanometers:

   _____

   (g) 200 nanometers:

   _____

   (h) 4×10^4 nanometers:

   _____

5. What colors of visible light are produced by the following wavelengths?
   (a) 700 nanometers:

   _____

(b) 400 nanometers:

_____

(c) 600 nanometers:

_____

(d) 550 nanometers:

_____

(e) 480 nanometers:

_____

(f) 590 nanometers:

_____

6. Convert each of the wavelengths in Question 5 above to angstroms.

(a) _____

(b) _____

(c) _____

(d) _____

(e) _____

(f) _____

Photoconductive Transducer Characteristics

Introduction

Photoconductive transducers are designed to produce changes in their electrical conductivity when variations of light energy occur. These devices are also called *photoresistive*, since their resistance varies in inverse proportion to their conductivity. The cadmium sulfide (CaS) cell, shown in Fig. 3-2A, is a common type of photoconductive cell. When exposed to varying intensities of visible light, the cadmium sulfide cell will change resistance. An increase in light energy falling onto its surface will increase the conductivity of the cell. The cell is highly sensitive to variations of light intensity. It is typically used in alarm and relay control systems.

The operational principle of a photoconductive cell is that, when light strikes its surface, valance electrons of the semiconductor material are released from their atomic bonds. When electrons are released, the resistance of the material decreases. We may also say that the material becomes more conductive. If the light intensity is increased, more electrons will be released and the material becomes more conductive. Resistance of photoconductive devices may range from several megohms in darkness to 50–100 Ω in fairly intense light.

Objective

In this activity you will observe the characteristics of a photoconductive cell. The photoconductive cell is a popular type of photoelectric transducer.

Equipment

Photoconductive cell: GE-X6
Variable ac power supply
Multifunction meter
Lamp: 60 W
Connecting wires

Procedure

1. Assemble the circuit shown in Fig. 3-2B. Connect the meter across the photoconductive cell so that

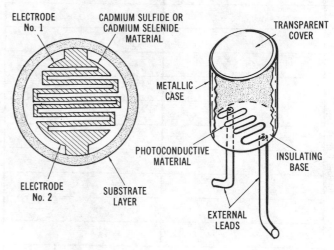

(a) Top view.   (b) Cutaway view.

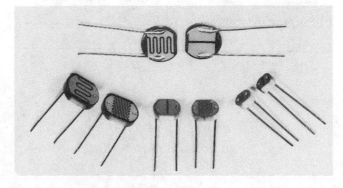

(c) Photograph.

Fig. 3-2A. Cadmium sulfide photoconductive cell.

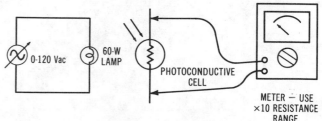

Fig. 3-2B. Photoconductive cell test circuit.

it will measure resistance changes with changes in light intensity. The 60-watt lamp is connected to a 0–120-Vac power source.

2. Record the resistance of the photoconductive cell with room light.

$R =$ _____ Ω

3. Cover the top of the cell with your finger and record the resistance without light.

$R = \text{_____} \ \Omega$

Table 3-2A. Photoconductive Cell Characteristics

Light Source Voltage	Resistance (Ω)
0	
10	
20	
30	
40	
50	
60	
70	
80	
90	
100	
110	
120	

4. Place the 60-watt lamp within ½ inch (1.27 cm) of the top of the cell. Adjust the variable ac source to zero volts.
5. Turn on the ac source and complete Table 3-2A by adjusting the source voltage as indicated.
6. This concludes the activity.

Analysis

1. What is the range of resistance change of the photoconductive cell used in this activity?

Range = _____ to _____ Ω
2. Discuss the characteristics of a photoconductive cell.

_____

_____

_____

_____

_____

_____

_____

Photoconductive Transducer Application

Introduction

One application of a photoconductive transducer is shown in Fig. 3-3A. In this circuit SCR_1 will conduct when light is focused onto the photoconductive cell. When light strikes the transducer, its resistance will decrease. The potential at point A becomes more positive and causes gate current (I_G) to flow. A sufficient amount of gate current will trigger the SCR into conduction. When SCR_1 conducts, the load device or output indicator will be activated. The SCR will then conduct until its anode circuit is opened. Variable resistor R_1 is used as a sensitivity adjustment to control the level of light required to cause the SCR to conduct. There are many similar applications of photoconductive transducers. This circuit could be modified to indicate changes in light intensity by replacing the SCR with some other type of control device.

Objective

In this activity you will construct a circuit which uses a photoconductive transducer to cause an output indicator to vary with changes in light intensity. By adjusting a potentiometer the sensitivity of the circuit may be varied. Two transistors amplify the dc variation caused by light focused onto the transducer. The circuit is used to control a current meter which could be calibrated to indicate changes in light intensity.

Equipment

Photoconductive cell (GE-X6 or equivalent)
Dc power supply

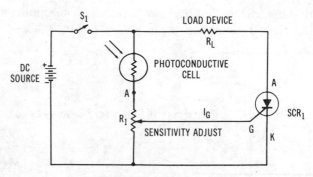

Fig. 3-3A. Photoconductive cell application.

Potentiometer: 50 kΩ
Variable ac power source
60-W lamp with socket
Transistors: 2N3053 (2)
Resistors: 2.7 kΩ, 20 kΩ
Electronic multifunction meter
Connecting wires

Procedure

1. Construct the circuit shown in Fig. 3-3B.
2. Place the light source about 6 inches (15.2 cm) from the photoconductive cell.
3. Adjust potentiometer R_1 while observing the output indicator. What occurs?
4. Pass a sheet of paper between the light source and the photoconductive cell. This should cause the output indicator to vary. If it does not, alter the light source and potentiometer R_1 to cause the desired reaction.
5. Complete Table 3-3A, first with the photoconductive cell exposed to light and then without light.

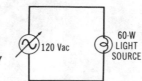

Fig. 3-3B. Output indicator controlled by photoconductive cell.

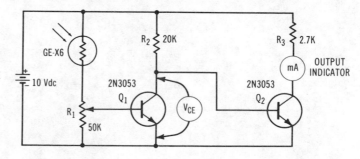

**Table 3-3A. Photoconductive Circuit Characteristics
(Forward Connection)**

Light Source	V_{CE} of Q_1	V_{CE} of Q_2	Output Indicator Current
With Light			
Without Light			

6. Alter the circuit by moving the photoconductive cell and potentiometer as shown in Fig. 3-3C.

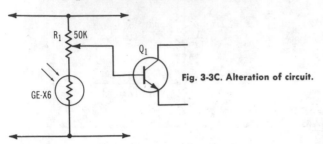

Fig. 3-3C. Alteration of circuit.

7. Alter the light source and the potentiometer until control of the output indicator is accomplished when the cell is exposed to light and then shielded from light.
8. When the circuit is working properly, complete Table 3-3B.

**Table 3-3B. Photoconductive Circuit Characteristics
(Reverse Connection)**

Light Source	V_{CE} of Q_1	V_{CE} of Q_2	Output Indicator Current
With Light			
Without Light			

9. This concludes the activity.

Analysis

1. How does the operation of the circuit of Fig. 3-3B compare to the operation of the circuit with the alteration of Fig. 3-3C?

_____

_____

_____

2. Write an analysis of the operation of the two circuits constructed in this activity.

_____

_____

_____

_____

_____

_____

_____

_____

3. Compare the data obtained in Tables 3-3A and 3-3B.

_____

_____

_____

_____

_____

Photoemissive Transducer Characteristics

Introduction

Phototubes are a common type of photoemissive transducer. The phototube, shown in Fig. 3-4A, is similar in appearance to other vacuum or gaseous tubes. This particular type of tube has a cathode which emits electrons when it is struck by light. This effect is called *photoelectric emission*. A phototube cathode emits electrons whose energy depends on the wavelength of the light striking it. A spectral response curve is plotted by the manufacturer for each type of phototube produced. A typical spectral response curve is shown in Fig. 3-4B.

Light energy exists in the form of photons, which are discrete bundles of nergy. When photoemission takes place in a phototube, photons of light are absorbed by the surface of the cathode. The absorption of these photons of light causes electrons on the surface of the cathode to gain enough energy to leave the cathode. The energy possessed by the electrons which leave the cathode is based on the frequency (or wave-

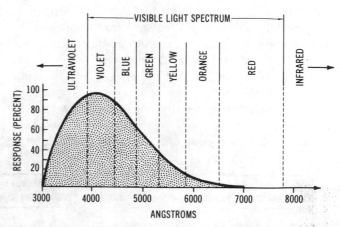

Fig. 3-4B. Spectral response curve for a phototube.

length) of the light. The amount of electrons which are emitted is based on the intensity of the light.

Objective

In this laboratory activity you will observe the characteristics of a photoemissive transducer. A phototube will be analyzed with light and without light. The phototube is a common type of photoemissive transducer.

Equipment

Variable dc power source
Variable ac power source
Lamp: 60 W
Phototube: 917
Resistor: 1 MΩ
Electronic multifunction meter
Cardboard cylinder

Procedure

1. Construct the circuit shown in Fig. 3-4C.
2. Place a cardboard cylinder over the phototube, with an open window facing the cathode of the tube.
3. Adjust the dc power source to produce 80 V. Adjust the distance and intensity of the light until

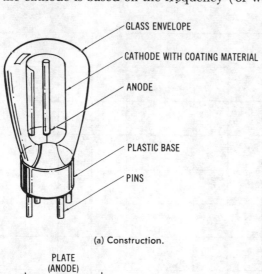

(a) Construction.

(b) Schematic symbols.

Fig. 3-4A. Phototube.

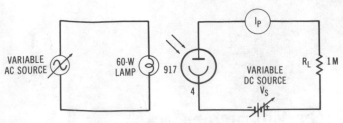

Fig. 3-4C. Phototube test circuit.

20 μA is indicated on the current meter connected in the plate circuit. Once the light has been positioned to achieve 20 μA, do not alter its position.

4. Complete Table 3-4A for the circuit of Fig. 3-4C with light applied to the phototube.
5. Complete Table 3-4B for the circuit with the light source turned off.
6. This concludes the activity.

Table 3-4A. Photocube Characteristics (With Light)

Source Voltage (V_S)	V_P	V_{RL}	I_P	Resistance* of Tube (Computed)
0 V				
5 V				
10 V				
20 V				
30 V				
40 V				
50 V				
60 V				
70 V				
80 V				

* $R = V_P/I_P$.

Analysis

1. Construct a plate voltage versus plate current graph with the data of Tables 3-4A and 3-4B that show the characteristics of the phototube with light and without light. Draw these two curves in Fig. 3-4D.
2. Discuss the operation of the phototube with light and without light.

_____

_____

_____

_____

_____

Table 3-4B. Phototube Characteristics (Without Light)

Source Voltage (V_S)	V_P	V_{RL}	I_P	Resistance of Tube (V_P/I_P)
0 V				
5 V				
10 V				
20 V				
30 V				
40 V				
50 V				
60 V				
70 V				
80 V				

_____

_____

3. As light intensity increases, what happens to: (a) V_P, (b) I_P, (c) V_{RL}, (d) tube resistance (R)?

(a) _____

(b) _____

(c) _____

(d) _____

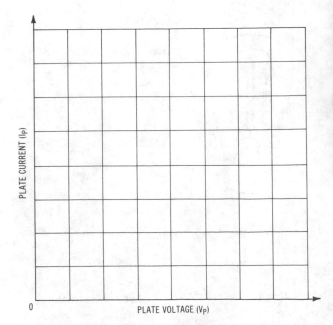

Fig. 3-4D. Observed phototube characteristics.

Photoemissive Transducer Applications

Introduction

There are many applications of photoemissive transducers. A phototube may be connected in a circuit as shown in Fig. 3-5A. When light is focused onto the cathode, the electrons which are emitted by the cathode travel to the positive potential of the plate.

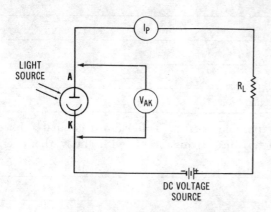

Fig. 3-5A. Phototube circuit action.

A plate current (I_P) will not flow, which causes a voltage drop across the load (R_L). The plate current, caused by various combinations of light and plate voltage, may be determined by using a phototube characteristic curve which is supplied by the manufacturer. Such a curve is shown in Fig. 3-5B. Photo-

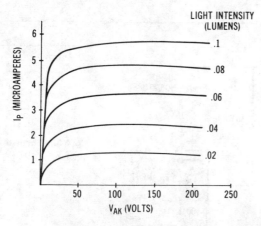

Fig. 3-5B. Phototube characteristic curves.

tubes may be either vacuum or gas-filled types. The gas phototube is more sensitive and thus requires less light to produce a given amount of anode current. A phototube circuit may be calibrated in order to measure light and be used in such instruments as light-exposure meters.

Objective

In this laboratory activity you will observe an application of a photoemissive transducer. The amount of light focused onto the phototube in this circuit controls the conduction of a vacuum-tube triode. By varying the grid potential of the triode, the variation of an output indicator can be accomplished. This is only one example of how a photoemissive transducer can be used in measurement.

Equipment

Phototube: 917 or equivalent
Dc power source
Variable ac power source
Vacuum tube: 6C4
Ac supply: 6.3 V
Lamp: 60 W
Resistors: 500 Ω (10 W), 5 kΩ (5 W), 10 kΩ, 12 kΩ, 33 kΩ, 5.6 MΩ
Connecting wires
Electronic multifunction meter

Procedure

1. Construct the circuit shown in Fig. 3-5C.
2. Adjust the intensity of the light source to maximum. Place the light about 6 inches from the phototube.
3. Pass a sheet of paper between the light source and the phototube. This should cause the output indicator to vary. If it does not, alter the light source to cause the desired reaction.
4. Make the measurements necessary to complete Table 3-5A, first with the phototube exposed to light and then without light.

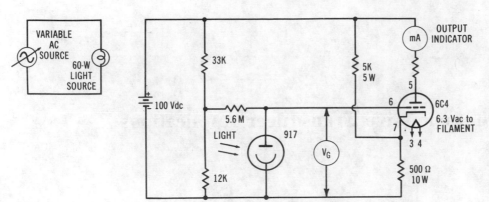

Fig. 3-5C. Output indicator controlled by phototube.

5. Alter the circuit by making the grid circuit changes shown in Fig. 3-5D.

Table 3-5A. Phototube Circuit Values (Forward Connection)

Light Source	Plate Voltage* (V_P)	Grid Voltage (V_G)	Output Indicator Current (mA)
With Light			
Without Light			

* Anode-to-cathode measurement.

6. Alter the light source until the output indicator varies when a sheet of paper is placed between the light and the phototube.
7. When the circuit is working properly, complete the measurements for Table 3-5B.
8. This concludes the activity.

Table 3-5B. Phototube Circuit Values (Reverse Connection)

Light Source	Plate Voltage (V_P)	Grid Voltage (V_G)	Output Indicator Current (mA)
With Light			
Without Light			

Analysis

1. How does the operation of the circuit of Fig. 3-5C compare with the operation of the circuit of Fig. 3-5D? Explain the difference.

_____

_____

_____

_____

2. From the data of Table 3-5A, compare the characteristics of this circuit with and without light.

_____

_____

_____

_____

3. From the data of Table 3-5B, compare the characteristics of this circuit with and without light.

_____

_____

_____

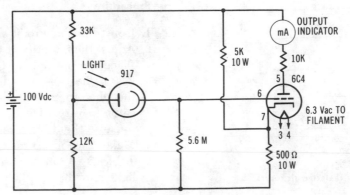

Fig. 3-5D. Modified phototube circuit.

4. What are some other applications of photoemissive
transducers?

_____

_____

_____

_____

_____

_____

_____

Photovoltaic Transducer Characteristics

Introduction

Photovoltaic transducers, commonly called *solar cells*, are used to convert light energy into electrical energy. Since this process is a direct energy conversion, much recent research has been conducted on how

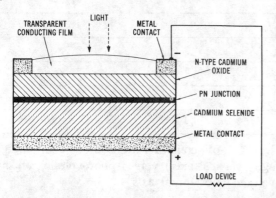

(a) Structure.

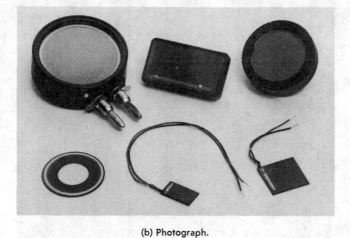

(b) Photograph.

Courtesy Vactec, Inc.

Fig. 3-6A. Selenium photovoltaic cells.

to convert large amounts of light energy into electrical energy. A common application of solar cells is in photographic light-exposure instruments. The electrical output of the solar cell is proportional to the amount of light falling onto its surface. The output is used to deflect a light-intensity meter movement.

The construction of a photovoltaic cell is shown in Fig. 3-6A. This selenium cell has a layer of selenium deposited on a metal base, and a layer of cadmium on another metal base. In the fabrication one layer of cadmium selenide and another layer of cadmium oxide is produced. A transparent conductive film is placed over the cadmium oxide, and a section of conductive alloy is then placed on the film. The external leads are connected to the conductive material around the cadmium oxide layer and the metal base. When light strikes the cadmium oxide layer, electrons are emitted and move toward the external load device. A deficiency of electrons is now created in this region, which is filled by electrons from the selenium material. Now electrons will be removed from the metal base into the selenium. Thus light energy causes a difference in potential to exist between the two external terminals.

Selenium cells have a very low efficiency of converting light energy to electrical energy; therefore silicon is now more often used since silicon cells have efficiencies as high as 15 percent. The more common silicon photovoltaic cell is shown in Fig. 3-6B. When no light is focused onto the silicon cell, it operates similar to a conventional pn junction diode. When light strikes the cell, a voltage is developed across the external leads. The more intense the light, the greater the potential difference across the cell.

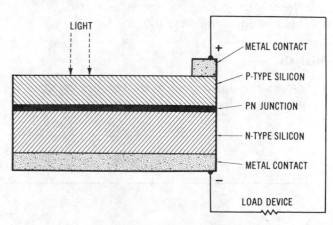

Fig. 3-6B. Silicon photovoltaic cell.

Objective

In this laboratory activity you will observe the characteristics of a photovoltaic transducer. You will notice that as the intensity of the light source is varied, the voltage output of the photovoltaic transducer varies proportionately.

Equipment

Electronic multifunction meter
Photovoltaic cell
Lamp: 60 W
Variable ac source

Procedure

1. Connect the voltmeter to the photovoltaic transducer (PV) as shown in Fig. 3-6C. Prepare the meter to measure a small dc voltage.

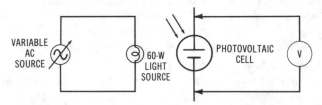

Fig. 3-6C. Photovoltaic transducer test circuit.

2. Record the dc voltage output generated by the photovoltaic cell in total darkness and in normal room light.

   Voltage in darkness = _____ Vdc

   Voltage in room light = _____ Vdc

3. Place a 60-W light bulb within ½ inch (1.27 cm) of the surface of the photovoltaic cell. Connect the bulb to the variable ac source. The source should be adjusted to zero.

4. Turn on the ac source and complete Table 3-6A by adjusting the source voltage as indicated.

5. This concludes the activity.

Analysis

1. Discuss the operational characteristics of a photovoltaic transducer.

   _____

   _____

   _____

Table 3-6A. Photovoltaic Transducer Characteristics

Light Source Voltage	Photovoltaic Cell Output Voltage

_____

_____

_____

_____

_____

2. In Fig. 3-6D construct a graph of light source voltage (vertical axis) versus photovoltaic cell output voltage (horizontal axis).

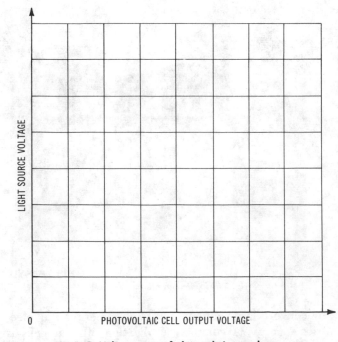

Fig. 3-6D. Voltage curve of photovoltaic transducer.

Photovoltaic Transducer Applications

Introduction

Photovoltaic transducers are used for a variety of applications. Although their electrical output is low, they may be used with amplifying devices to develop an output which will control an output indicator or some other load device. One such application is shown in Fig. 3-7A. In this circuit the output of the photovoltaic transducer is amplified by transistor Q_1, causing an increase in base current (I_B) when light strikes the photovoltaic cell. This increased base current causes the collector current (I_C) of the transistor to increase also. The collector current will be of sufficient value to cause the output indicator to respond.

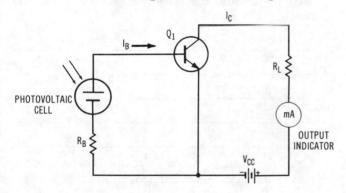

Fig. 3-7A. Photovoltaic cell application.

Objective

In this laboratory activity you will construct two circuits which illustrate applications of a photovoltaic transducer. In each circuit a photovoltaic cell will be used to control the conduction of a field-effect transistor (FET). The second circuit shows how an output indicator will vary with changes of light intensity focused onto a photovoltaic cell.

Equipment

Photovoltaic cell (solar cell)
Field-effect transistor: GE-FET-1
Resistors: 560 Ω, 1 kΩ, 10 kΩ
Electronic multifunction meter

Variable dc power supply
Transistor: 2N3053
Lamp and socket: 60 W
Variable ac power source
Assorted colors of thin opaque plastic sheets
Connecting wires

Procedure

1. Construct the circuit shown in Fig. 3-7B.

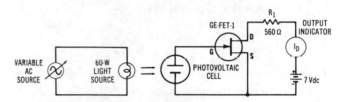

Fig. 3-7B. FET circuit controlled by photovoltaic cell.

2. Record the gate voltage (V_G), drain current (I_D), and source-drain voltage (V_{SD}), with the photovoltaic cell in darkness and in normal room light.

Darkness	Room Light
$V_G =$ _____ Vdc	$V_G =$ _____ Vdc
$I_D =$ _____ mA	$I_D =$ _____ mA
$V_{SD} =$ _____ Vdc	$V_{SD} =$ _____ Vdc

3. Place the 60-W lamp about ½ inch (1.27 cm) from the photovoltaic cell. Complete Table 3-7A.
4. Construct the circuit shown in Fig. 3-7C and check it for operation.
5. Record the drain current (I_D) of Q_1 and the collector current (I_C) of Q_2 with the cell in total darkness and normal room light.

Darkness	Room Light
I_D of $Q_1 =$ _____ mA	I_D of $Q_1 =$ _____ mA
I_C of $Q_2 =$ _____ mA	I_C of $Q_2 =$ _____ mA

6. Record the output indicator current (I_D) when an opaque white surface is placed 1 inch (2.54 cm) from the photocell in normal room light.

Current = _____ mA

7. Record the output current when three opaque surfaces of different colors are placed, one at a time, 1 inch (2.54 cm) from the photocell in normal room light. Record each color in the blanks below.

Colors	Output Current
No. 1 _____	_____ mA
No. 2 _____	_____ mA
No. 3 _____	_____ mA

8. Record the output current for two different opaque surfaces that seem to be of identical colors.

Color No. 1 _____; current = _____ mA

Color No. 2 _____; current = _____ mA

9. This concludes the activity.

Analysis

1. Analyze the operation of the circuits that appear in Figs. 3-7B and 3-7C.

_____

_____

_____

_____

_____

_____

Table 3-7A. Photovoltaic Cell Circuit Measurements

Light Source Voltage	V_G	V_{SD}	I_D
0 V			
10 V			
20 V			
30 V			
40 V			
50 V			
60 V			
70 V			
80 V			
90 V			
100 V			
110 V			
120 V			

2. Describe how the circuit of Fig. 3-7C could be used to measure the thickness of translucent materials.

_____

_____

_____

_____

3. From the data of Table 3-7A, as light source intensity increases, what happens to the following: (a) Gate voltage (V_G)?

_____

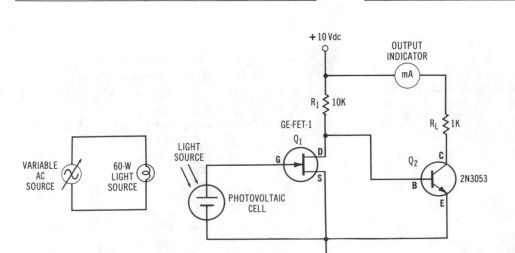

Fig. 3-7C. Output indicator controlled by light.

(b) FET source-drain voltage (V_{SD})?

_____

(c) Drain current (I_D)?

_____

4. Compare the current measurements recorded in Step 5 for dark and light conditions.

_____

_____

5. What do the measurements of Steps 6 and 7 illustrate?

_____

_____

_____

Light-Emitting Diode (LED) Characteristics

Introduction

In Unit 1 you studied instrument readout devices which utilized light-emitting diodes (LEDs). LEDs have become a very important type of photoelectric device. You should become familiar with their characteristics.

The light-emitting diode is also called a *solid-state lamp*. Fig. 3-8A shows an LED. This device is small and lightweight, making it desirable for use with digital circuitry and other miniaturized applications. The pn junctions of LEDs are commonly made of gallium arsenide phosphide (GaAsP) and gallium phosphide (GaP). These elements provide rapid response time and cause the LED to present a low impedance compatible with solid-state circuitry.

The LED is operated in a forward bias condition. The semiconductors used in its fabrication have the unique property of producing photoemission when a

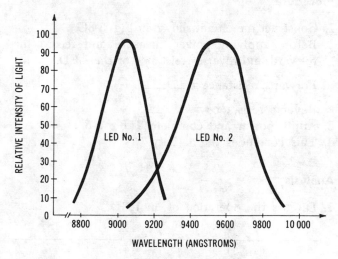

Fig. 3-8B. LED spectral response curves.

forward bias potential is applied. Thus electrical energy causes the radiation of visible light energy. The spectral response of LEDs varies according to the type of semiconductor materials used in their design. LEDs are made to produce different colors. Typical LED spectral response curves are shown in Fig. 3-8B. The wavelength of radiated energy from an LED is beyond the visible range. Therefore phosphors are added to produce the desired color effects. LEDs are sensitive to temperature changes. Any change in temperature will cause the spectral response to shift.

Objective

In this laboratory activity you will observe the characteristics of a light-emitting diode (LED). LEDs are used for numerous applications in instrumentation and elsewhere.

Equipment

Light-emitting diode
Electronic multifunction meter
0–10-Vdc power source
1-kΩ resistor
Connecting wires

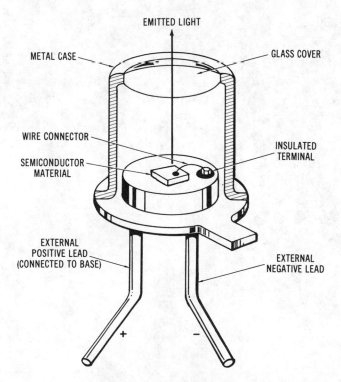

Fig. 3-8A. Light-emitting diode.

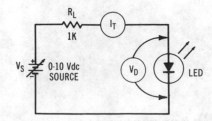

Fig. 3-8C. LED circuit.

Procedure

1. Construct the circuit shown in Fig. 3-8C.
2. Before applying power, measure and record the forward and reverse resistance of the LED.

   Forward resistance = _____

   Reverse resistance = _____
3. Apply power and complete Table 3-8A.
4. This concludes the activity.

Analysis

1. Discuss the operation of an LED.

_____

_____

_____

Table 3-8A. LED Characteristics

V_S	V_{RL}	V_D	I_T	$R_D = V_D/I_T$	Light Emitted*
1 V					
2 V					
3 V					
4 V					
5 V					
6 V					
7 V					
8 V					
9 V					
10 V					

* Light as detected by the human eye, i.e., dim, moderate, bright.

_____

_____

_____

_____

_____

_____

Photoelectric Pulse Modulation and Demodulation

Introduction

An important application of photoelectric transducers is in modulation and demodulation circuits. Modulation is the process of converting dc to ac, and demodulation is the conversion of ac to dc. A light source can be used to change the electrical characteristics of a photoelectric transducer. The type of input applied to the light source determines the electrical operation of a photoelectric transducer located adjacent to the source. Light pulses of varying frequency can be used to cause output variation of a photoelectric transducer. Circuits that perform the functions described above are called *chopper circuits* and are used in many instruments, such as electronic voltmeters.

Objective

In this activity you will construct a photoelectric chopper circuit and observe its operating characteristics with a variable-frequency signal applied to its input. An LED is used as the light source and a phototransistor circuit is used to sense changes in light.

Equipment

Light-emitting diode
Audio signal generator
Phototransistor: GE-X19
Electronic multifunction meter
Dc power source
Resistors: 1 kΩ (2), 100 kΩ

Potentiometer: 50 kΩ
Connecting wires

Procedure

1. Construct the circuit shown in Fig. 3-9A.
2. Place the LED so that it touches the phototransistor.
3. Apply power to the phototransistor circuit.
4. Adjust I_C to about 3 mA. Then measure and record I_B and V_B.

   $I_B =$ _____ mA

   $V_B =$ _____ V

5. Cover the LED and phototransistor so that they are isolated from external light.
6. Turn on the audio signal generator and adjust its output voltage to 6 V rms. Set the frequency to 50 Hz.
7. Complete Table 3-9A by varying the signal generator frequency and measuring the values of the phototransistor circuit.

Table 3-9A. Photoelectric Chopper Circuit Values

Frequency	V_B	I_B	I_C
500 Hz			
1 kHz			
2.5 kHz			
5 kHz			
10 kHz			
20 kHz			

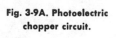

Fig. 3-9A. Photoelectric chopper circuit.

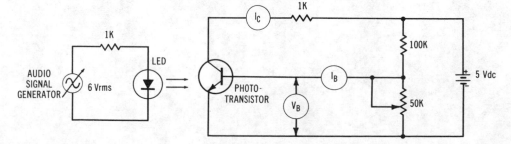

8. This concludes the activity.

Analysis

1. As frequency of the signal generator increases, what happens to: (a) I_B, (b) V_B, (c) I_C?

   (a) _____

   (b) _____

   (c) _____

2. How does this circuit act as a chopper?

   _____

3. How does the LED cause phototransistor operation to change?

   _____

   _____

   _____

   _____

Hartley Oscillator Pulse Modulator

Introduction

When light is used to transmit information concerning a quantity being measured, it is sometimes essential that the transmitted light exhibit certain characteristics. These light characteristics aid in allowing the receiving mechanism to distinguish between the light that represents information and room light. One method of creating a light source with a very high pulse per second (pps) rate employs the Hartley oscillator and a solid-state lamp.

The Hartley oscillator is a sine-wave oscillator that generates an alternating signal (Fig. 3-10A). Signal frequency is controlled by the values of L and C in the tank circuit while signal amplitude is controlled by feedback and the biasing network of the amplifier.

When a solid-state lamp (LED) is connected across the tank circuit of the oscillator, it emits a pulsating light. Its pps rate is the same as the frequency of the waveform generated by the oscillator.

Objective

In this activity you will construct a Hartley oscillator circuit and examine the oscillator output under various conditions.

Equipment

Digital vom
Oscilloscope
Light-emitting diode
Transistor: 2N2405
Coils: Center-tapped oscillator coil, 30-mH choke coils
Capacitors: 47 pF; 0.01 μF, 50 Vdcw
Variable capacitor: 100 pF
Resistors: 2 kΩ; 15 kΩ
Dc power supply
Connecting wires

Procedure

1. Construct the circuit shown in Fig. 3-10B.
2. Measure and record the resistance of L_1 from the center tap to each end.

   CT to top $R =$ _____ Ω

   CT to bottom $R =$ _____ Ω
3. Connect the high-resistance side of the coil to C_1 and the low-resistance side to C_2.
4. Connect the oscilloscope at points A and B and draw the waveforms displayed by the scope below.

Fig. 3-10A. Transistor Hartley oscillator.

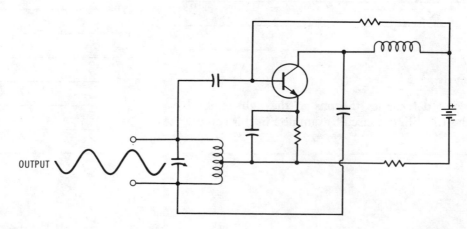

OUTPUT

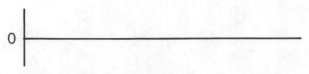

5. While observing the scope, adjust C_3 through its entire range. How does this affect the frequency of the oscillator?

_____

_____

_____

6. While observing the scope, remove the connecting wire from C_2 to the tank circuit. How does this affect the action of the oscillator? Why?

_____

_____

_____

7. Return the connecting wire removed in Step 6 and connect the LED to points A and B.
8. Draw the waveforms that appear across the LED.

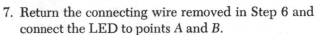

9. How did connecting the LED affect the waveform across the tank circuit?

_____

_____

_____

10. How do the waveforms in Step 4 compare with those in Step 8?

_____

_____

_____

Analysis

1. What determines the pps of the light emitted by the LED in the circuit examined in this activity?

_____

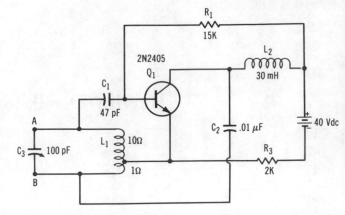

Fig. 3-10B. Test Hartley oscillator.

_____

2. How can the pps of the light be altered?

_____

_____

3. Where might a light modulator of the type examined in this activity be used?

_____

_____

_____

4. Write an analysis of the operation of the Hartley oscillator.

_____

_____

_____

_____

_____

_____

_____

_____

_____

_____

Astable Multivibrator Pulse Modulator

Introduction

The astable multivibrator is a special type of free-running relaxation oscillator that generates a square-wave signal. (See Fig. 3-11A.) The frequency of the signal generated by this oscillator is determined by the values of C_1, R_2, C_2, and R_3, while the amplitude of the signal is determined by the values of R_2, R_3, Q_1, Q_2 and the dc voltage source.

This oscillator is no more than two amplifiers, each controlling the other's conductivity. When Q_1 is conducting, usually at the saturation level, Q_2 is off. As the conduction level of Q_1 drops below saturation, Q_2 begins conduction and rises very quickly to saturation as Q_1 is turned off. This cycle is repeated as long as the circuit is energized.

When a solid-state lamp (LED) is connected in such a way as to be energized by the oscillator, a very excellent pulse modulator is created. The pps rate of the emitted light is the same as the frequency of the oscillator.

Objective

In this activity you will construct an astable multivibrator pulse modulator and observe the effect on it of changing resistance and capacitance values.

Equipment

Oscilloscope
Light-emitting diode
Transistors: 2N2405 (2)

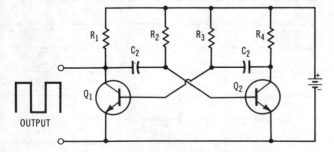

Fig. 3-11A. Astable multivibrator.

Resistors: 100 Ω (3), 2.7 kΩ (2), 4.7 kΩ
Capacitors: 0.01 μF (2), .02 μF, 50 Vdcw
Dc power supply
Connecting wires

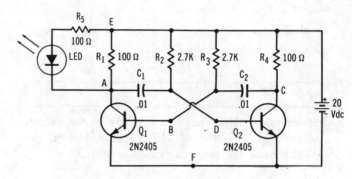

Fig. 3-11B. Astable multivibrator pulse modulator.

Procedure

1. Construct the circuit in Fig. 3-11B.
2. Record the waveform between points A and E of the circuit.

   Waveform at A-E _____

3. Connect the vertical side of the scope at points A, B, C, and D in relation to point F and record the waveforms in the space below. (NOTE: The scope should be adjusted to produce two to four waveforms.)

   Waveform at A-F _____

   Waveform at B-F _____

   Waveform at C-F _____

   Waveform at D-F _____

4. Replace C_1 with a 0.02-μF capacitor and record the waveforms at the same points as in Step 3.

Waveform at A-F _____

Waveform at B-F _____

Waveform at C-F _____

Waveform at D-F _____

5. How do the waveforms recorded in Step 3 compare with those in Step 4?

_____

_____

6. Return the 0.01-μF capacitor as C_1 to the circuit and replace R_2 with a 4.7-kΩ resistor. Record the waveforms at the same points as in Step 3.

Waveform at A-F _____

Waveform at B-F _____

Waveform at C-F _____

Waveform at D-F _____

7. How do the waveforms recorded in Step 3 compare with those in Step 6?

_____

_____

8. Return the 2.7-kΩ resistor as R_2 to the circuit and connect the scope across the LED. Record the waveform in the space below.

Waveform across LED _____

Analysis

1. What is meant by the terms *pulses per second* (pps)?

_____

_____

_____

2. Why is this oscillator termed *astable*?

_____

_____

3. What determines the frequency of the oscillator?

_____

_____

4. What determines the pps rate of the light emitted by the LED?

_____

_____

5. Write an analysis of the operation of the circuit illustrated in Fig. 3-11B.

_____

_____

_____

_____

_____

_____

_____

_____

_____

_____

_____

_____

_____

Pulse Modulator Detector

Introduction

Most of the time, when a pulse modulator is used as a part of a system, a pulse modulator detector and amplifier combination is also necessary.

The pulse modulator detector is a circuit designed to respond to the pulses of light generated by a pulse modulator. It will not respond to outside or ambient light; therefore it "receives" only the light messages transmitted by the pulse modulator and is isolated from outside interference.

The amplifier normally used with the detector is usually a high-frequency amplifier designed to best respond to an electrical signal whose frequency is that of the pulses of light generated by the pulse modulator. The readout device is normally connected to the output of the amplifier.

Objective

In this activity you will have the opportunity to examine the operating characteristics of a pulse modulator with its amplifier.

Equipment

Oscilloscope
Digital vom
Signal generator
Light-emitting diode
Transistors: Phototransistor, 2N1086

Capacitors: 50 μF, 20 μF, 0.1 μF, 50 Vdcw
Resistors: 4.7 kΩ, 120 kΩ, 1 MΩ
Output transformer, 2.5-H primary
Dc power supply
Connecting wires

Procedure

1. Construct the circuit illustrated in Fig. 3-12A.
2. Record the voltage output as indicated by the voltmeter of the circuit.

   Output = _____ V

3. Adjust the signal generator to produce a signal frequency above and below 300 Hz and describe the effect upon the output as measured by the voltmeter of the circuit.

   _____

   _____

   _____

4. Return the signal generator to 300 Hz.
5. Pass a piece of paper between Q_1 and the light of the room and describe the effect upon the output of the circuit.

   _____

   _____

   _____

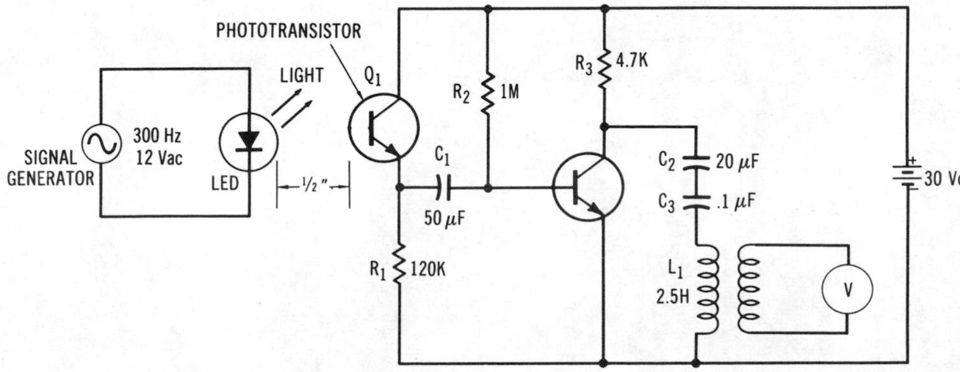

Fig. 3-12A. Pulse modulator detector.

153

6. Pass a piece of paper between Q_1 and the LED and describe the effect upon the output of the circuit.

_____

_____

_____

7. How do the results of Steps 5 and 6 compare?

_____

_____

_____

8. Connect an oscilloscope across the LED and record the waveforms in the space below.

0

9. Using the data gathered in Step 8, draw the characteristics of the light pulses being emitted by the LED.

0

10. Where might the pulse modulator detector be used in an industrial measurement system?

_____

_____

_____

Analysis

1. What is the purpose of a pulse modulator detector?

_____

_____

2. How can the pulse modulator detector respond to light that is pulsating at a specific frequency, while rejecting all other types of light?

_____

_____

_____

3. What is the purpose of the amplifier portion of the pulse modulator detector?

_____

_____

_____

4. Write an analysis of the operation of the pulse modulator detector examined in this activity.

_____

_____

_____

_____

_____

_____

_____

_____

_____

_____

Miscellaneous Optical Coupling

Introduction

Often it is essential that a quantity being measured be changed to an electrical signal that is, in turn, changed to a medium easily transmitted, like light. The block diagram in Fig. 3-13A illustrates how this might be true for a temperature measurement system.

Previously, you have examined circuits and devices that changed temperature values and light into electrical signals. By using a combination of these devices, with appropriate electrical circuitry, we can convert temperature values into light, transmit the light, and convert the transmitted light back into temperature readings.

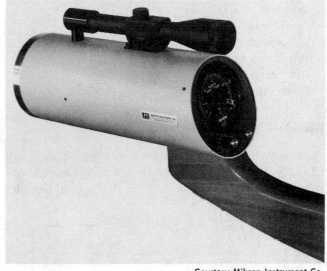

Courtesy Mikron Instrument Co.

Infrared thermometer with scope and rifle stock for sighting stability.

In the open system, light is transmitted in open space, from one point to another, and becomes part of the surrounding or normal environmental light. In the closed system, light is transmitted in a carefully sealed container, chamber, or cylinder and does not become part of the surrounding light. A typical closed optical coupling system is shown in Fig. 3-13C.

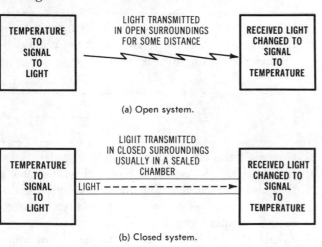

(a) Open system.

(b) Closed system.

Fig. 3-13B. Basic coupling systems.

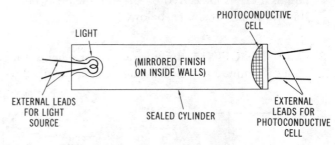

Fig. 3-13C. Closed optical coupling system.

Essential to a system of this nature is the method of optical coupling used in transmitting and receiving the light messages. Two basic optical coupling systems are used for this purpose. These are the open and closed systems as shown in Fig. 3-13B.

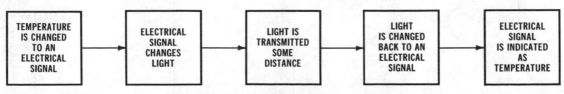

Fig. 3-13A. Temperature measurement system using light.

Objective

In this activity you will examine a closed optical coupling system.

Equipment

Digital vom
Photomod optical coupling system by Clairey (CLM7H16A073)
Thermistor: GE-X15
Resistors: 10 kΩ (2)
Potentiometer: 50 kΩ
Dc power supplies (2)
Styrofoam cup
Hot water
Connecting wires

Procedure

1. Construct the circuit shown in Fig. 3-13D.
2. Adjust the dc power supply until the resistance of the photoconductive cell of the optical coupler

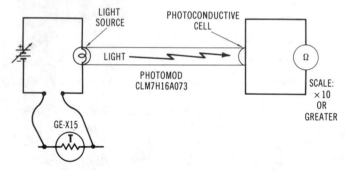

Fig. 3-13D. Thermistor-controlled optical coupling circuit.

equals 5 kΩ as measured by the ohmmeter. Record this value of resistance below.

$R =$ _____ Ω

3. Grasp the GE-X15 thermistor in your hand for about 3 minutes. Record the resistance of the

photoconductive cell caused when the thermistor was grasped.

$R =$ _____ Ω

4. Allow the thermistor to return to room temperature.
5. Immerse the thermistor in hot water (120°F–130°F or 48°C–54°C) for a period of 3 minutes. Record the resistance of the photoconductive cell after this action.

$R =$ _____ Ω

6. How did the resistance of the photoconductive cell recorded in Step 2 compare with that recorded in Step 3?

_____

_____

7. How did the resistance recorded in Step 2 compare with Step 5?

_____

_____

8. What is the relation between the temperature of the thermistor and the resistance of the photoconductive cell in the optical coupler?

_____

_____

9. Construct the circuit illustrated in Fig. 3-13E.
10. In Fig. 3-13E R_2 could be replaced with either a thermistor or an rtd. The resistance of both devices will increase or decrease with their environmental temperatures.
11. Adjust R_2 slowly through its range (from minimum to maximum resistance) and describe how this affects the collector current of Q_2.

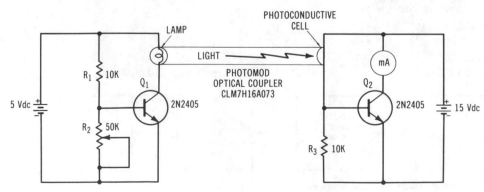

Fig. 3-13E. Two-transistor coupling circuit.

_____

_____

_____

12. Record the minimum and maximum collector current of Q_2 that is caused by the changing resistance of R_2.

Minimum R_2 causes _____ mA

Maximum R_2 causes _____ mA

Analysis

1. Describe how temperature values can be changed to light.

_____

_____

_____

2. Describe how light may be transmitted by either an open or closed system.

_____

_____

_____

_____

3. Describe how light can be converted into temperature values.

4. What are the advantages and disadvantages of both the open and closed optical coupling systems?

_____

_____

_____

_____

_____

5. Write an analysis of the operation of the optical coupling circuit illustrated in Fig. 3-13E.

_____

_____

_____

_____

_____

_____

_____

_____

_____

_____

Miscellaneous Input Transducers for Instrumentation Systems

There are many types of transducers used in industry to measure such quantities as pressure, force, weight, position, and displacement. Of these specialized transducers, the most widely used are the displacement transducers and strain gages.

The displacement transducers are those used to monitor, detect, or measure movement, position, or displacement. These transducers achieve their purpose through potentiometric, inductive, capacitive, or magnetic action.

Potentiometric transducers sense, detect, or measure displacement through a change in resistance brought about by a displacement. Usually the slider of a potentiometric device is mechanically coupled or linked to the object whose displacement is being monitored. Any movement or displacement of the subject results in a change in resistance of the potentiometer.

Inductive displacement transducers measure displacement when a movable core changes the inductance of a coil. Generally the device whose position is being monitored is mechanically linked to the movable core of an inductor. As the object being monitored is displaced, the core of the inductor is moved and the induction of the core is proportionally changed.

Capacitive displacement transducers depend upon a changing spatial relationship in capacitive plates when the transducer is monitoring the position of an object. The capacitance is altered when the distance between the plates of the capacitor is changed or when the exposed plate area is changed. When these plates are linked or coupled to a monitored object, any change in the position of the object brings about a change in the capacitance of the transducer.

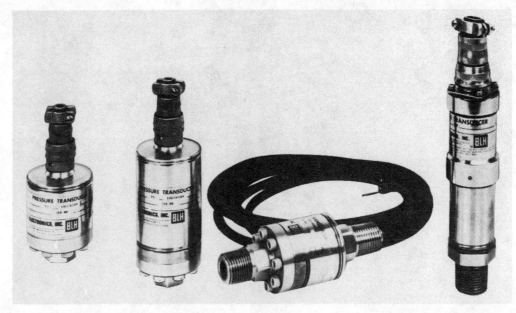

Pressure transducers.

Courtesy BLH Electronics, Inc.

One of the most popular magnetic displacement transducers is the linear variable-differential transformer (lvdt). This transformer is constructed with one primary coil magnetically coupled to two secondary coils through a movable core. As the core is displaced, more voltage is induced into one of the secondary coils than the other. Since the two secondary coils may be connected in series-opposing, the result is a voltage output indicative of the movement of the core. As with many of the other displacement transducers, the movable core is coupled to the object whose displacement is being monitored. Any change in the position of the monitored object results in a lvdt voltage output.

In this unit you will have an opportunity to examine many of the characteristics of the previously mentioned displacement transducers.

Displacement transducers.

Courtesy BLH Electronics, Inc.

Potentiometric Displacement Transducers

Introduction

Displacement transducers are those devices that detect a deviation in position, alignment, or location. This deviation becomes an electrical signal whose polarity indicates the direction of displacement. The magnitude of the electrical signal represents the degree or amount of displacement.

The potentiometric displacement transducer is the simplest and most easily understood of all displacement transducers. Generally the wiper of a potentiometer is mechanically linked to the object whose displacement is being monitored or detected. Fig. 4-1A shows a very simple potentiometric displacement transducer. If the object being monitored is displaced or moved a fraction of an inch on the horizontal plane, this movement will change the position of the wiper on the potentiometer. Changing the position of the wiper will, in turn, cause the circuit resistance of the potentiometer to be linearly altered. This will cause more or less current to flow through the milliammeter. If the milliammeter were calibrated relative to movement or deviation in centimeters or decimals, then this very simple circuit could be used to measure displacement.

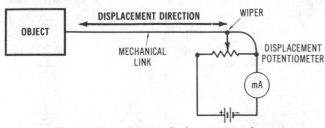

Fig. 4-1A. Potentiometric displacement transducer.

A variation of the previously illustrated circuit would allow displacement to be converted to voltage rather than current. In Fig. 4-1B any vertical movement of the object being monitored would readily appear as a change in voltage drop across R_1.

Potentiometric displacement transducers can also be used to measure liquid or gas pressure and liquid level when the proper mechanical coupling is employed.

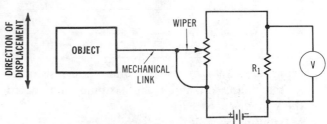

Fig. 4-1B. Circuit converting displacement to voltage.

Objective

In this activity you will observe the characteristics of the potentiometer displacement transducer.

Equipment

Digital vom
Dc power supply
Resistors: 1 kΩ (2), 150 Ω
Variable resistor with slider: 250 Ω, 5 W
Connecting wires

Procedure

1. The potentiometer or variable resistor used in this activity to represent a potentiometric displacement transducer is illustrated in Fig. 4-1C.

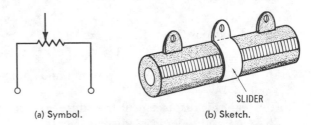

(a) Symbol.   (b) Sketch.

Fig. 4-1C. Potentiometric resistor.

2. Construct the circuit in Fig. 4-1D.
3. Assume that the slider of the potentiometric resistor is mechanically linked to an object whose displacement is being monitored. Move the slider slowly along the body of the potentiometric resistor and describe the effect this action has upon the voltage across R_1.

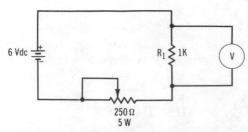

Fig. 4-1D. Basic potentiometric transducer circuit.

_____

_____

_____

4. Move the slider of the potentiometric resistor to its maximum left and right locations and record the voltages across R_1.

Maximum left position = _____ V

Maximum right position = _____ V

5. Construct the circuit shown in Fig. 4-1E.

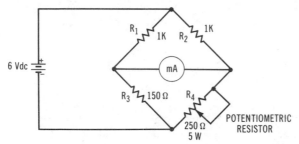

Fig. 4-1E. Potentiometric transducer bridge circuit.

6. Move the slider of the potentiometric resistor to the maximum left- and right-hand position and record the current along with the correct polarity displayed by the milliammeter (dvom).

Left-hand position = _____ mA;

polarity _____
Right-hand position = _____ mA;

polarity _____

7. Move the slider to the null position and complete Table 4-1A.

8. How can this displacement transducer detect both direction and amount of displacement?

_____

_____

_____

_____

9. Which of the two circuits used in Steps 2 and 5 is the more sensitive?

Table 4-1A. Potentiometric Transducer Bridge Circuit Data

Direction of Displacement from Null	Distance of Displacement Inches	Centimeters	Current (Milliamperes)	Polarity (+ or −)
Left	1/16	0.159		
Left	1/8	0.318		
Left	3/16	0.477		
Left	1/4	0.635		
Left	5/16	0.793		
Left	3/8	0.953		
Left	1/2	1.270		
Left	3/4	1.905		
Return to Null Position				
Right	1/16	0.159		
Right	1/8	0.318		
Right	3/16	0.477		
Right	1/4	0.635		
Right	5/16	0.793		
Right	3/8	0.953		
Right	1/2	1.27		
Right	3/4	1.91		

_____

_____

10. How could the action of the potentiometric displacement transducer be used to control the conductivity of an amplification circuit?

_____

_____

_____

Analysis

1. What are displacement transducers?

_____

_____

2. Where are displacement transducers used?

_____

_____

_____

3. How could a potentiometer displacement transducer be used to measure liquid or gas pressure?

_____

_____

_____

4. What is the purpose of the mechanical linkage between the object whose displacement is being monitored and the potentiometric transducer?

_____

_____

_____

Capacitive Displacement Transducers

Introduction

The capacitive displacement transducer detects and measures the displacement of an object to which it is linked due to the relationship between or among its plates.

The capacitance of any two charged areas or plates separated by a dielectric is determined by plate area, dielectric material, and distance between the plates. Causing the plates of any capacitor to be moved apart will result in a decreased capacitance (Fig. 4-2A). Decreasing the distance between the plates of a capacitor results in an increased capacitance.

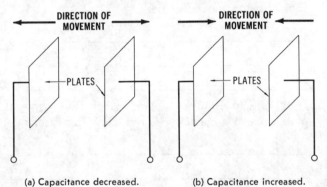

(a) Capacitance decreased.   (b) Capacitance increased.

Fig. 4-2A. Moving capacitor plates apart and together.

Likewise, by causing the exposed plate area to increase or decrease, capacitance is altered (Fig. 4-2B).

Altering the capacitance of a capacitor will cause capacitative reactance (X_C) to change. Increasing capacitance results in a decrease in X_C while decreas-

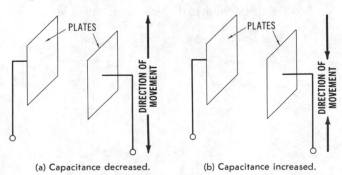

(a) Capacitance decreased.   (b) Capacitance increased.

Fig. 4-2B. Decreasing and increasing exposed plate area.

ing capacitance causes X_C to increase. Employing this principle with an ac current of a specified frequency readily converts the displacement of the plates of a capacitor to an electrical signal. When the capacitor plates are mechanically linked to an object whose displacement is being monitored, a very excellent displacement transducer is created.

Objective

In this activity you will examine the characteristics of the capacitive displacement transducer.

Equipment

Digital vom
Signal generator
Resistors: 10 kΩ, 100 kΩ, 25 kΩ
Fixed capacitor: 47 pF, 50 Vdcw
Variable capacitor: 20–150 pF, 50 Vdcw
Potentiometer: 500 kΩ
Connecting wires

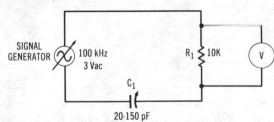

Fig. 4-2C. Basic capacitive displacement circuit.

Procedure

1. Construct the circuit shown in Fig. 4-2C.
2. Adjust C_1 to its maximum clockwise and counterclockwise positions and record the voltages across R_1.

   Ccw position _____ V

   Cw position _____ V
3. How does adjusting C_1 affect the voltage across R_1?

_____

_____

4. Construct the circuit in Fig. 4-2D.
5. Adjust C_2 and R_3 until the null state is achieved.
6. Assume the shaft of C_2 was mechanically linked to an object whose displacement was being monitored. By adjusting the shaft of C_2 manually, you will represent the monitored object being displaced. Adjust C_2 as prescribed in Table 4-2A and complete the table.

Table 4-2A. Simulation of Capacitive Displacement Transducer

| Amount of Displacement from Null (Adjustment of C_2) | | Resulting Voltage Across R_2 |
Inches	Centimeters	
1/16	0.159	
1/8	0.318	
3/16	0.477	
1/4	0.635	
5/16	0.793	
3/8	0.953	
7/16	1.111	
Maximum		

7. Describe the effect of object displacement upon the voltage across R_2.

_____

_____

_____

8. Record the voltage across R_2 when the capacitor is adjusted to represent both minimum and maximum displacement.

Maximum displacement = _____ V

Minimum displacement = _____ V

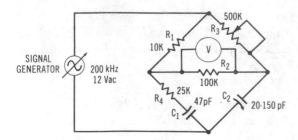

Fig. 4-2D. Capacitive displacement transducer bridge circuit.

9. Which of the circuits illustrated in Figs. 4-2C and 4-2D is the more sensitive?

_____

10. How could the circuit illustrated in Step 4 be altered to indicate direction as well as amount of displacement?

_____

_____

_____

Analysis

1. What is the effect upon capacitance when the plates of a capacitor are moved closer together?

_____

_____

2. How is capacitance affected when the plates of a capacitor are moved apart?

_____

_____

3. How is capacitance affected when the exposed plate area of a capacitor is reduced? Increased?

_____

_____

4. What is the relation between capacitance and X_C?

_____

_____

_____

5. Could a capacitance displacement transducer be used with direct current?

_____

_____

_____

_____

_____

Inductive Displacement Transducers

Introduction

The action of the inductive displacement transducer relies upon controlling the inductance of an inductor. The inductance of any inductor is determined by the number of turns of wire that makes up its coil and by the core material. Since the number of turns of its coil are difficult to alter, inductance is changed by altering the core material.

Generally, there are two broad classifications of inductors relative to core material. These are iron-core and air-core coils. For any given number of turns of coil wire, an inductor with an iron core will exhibit a greater inductance than one with an air core. Therefore, by simply altering the core material of an inductor (causing it to become more iron than air or vice versa), we can alter its inductance. See Fig. 4-3A.

By altering an inductor core, its inductance and inductive reactance (X_L) can be controlled. Increasing the iron content of an inductor core increases inductance and results in a proportional decrease in X_L. Decreasing the iron content of an inductor core causes the opposite effect. Since X_L is a form of ac impedance, displacement of an inductor core is easily converted to an electrical signal.

Fig. 4-3B shows an inductive displacement transducer mechanically linked to the object being monitored. Any displacement of the object on the horizontal plane results in a change in the voltage across R_1.

Objective

In this activity you will observe the characteristics of an inductive displacement transducer.

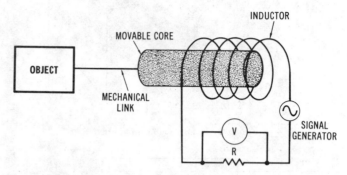

Fig. 4-3B. Inductive displacement transducer.

Equipment

Digital vom
Signal generator
Air-core coils, 100 turns, A.S. No. 16 wire (2)
6-inch (15-cm) steel cores, ¾-inch (1.91-cm) diameter (2)
Resistor: 10 kΩ
Potentiometer: 500 kΩ
Connecting wires

Procedure

1. Construct the circuit illustrated in Fig. 4-3C by using the coil as shown.
2. Position the movable core as nearly in the center of the coil as possible and record the voltage across R_1.

   Voltage across $R_1 = $ _____ V
3. Slowly move the core inside the coil from the left to right and right to left and record the minimum and maximum voltage across R_1 that is caused by this action.

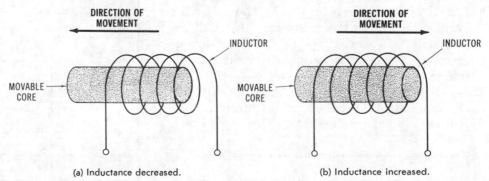

Fig. 4-3A. Movable core can increase or decrease inductance.

(a) Inductance decreased.

(b) Inductance increased.

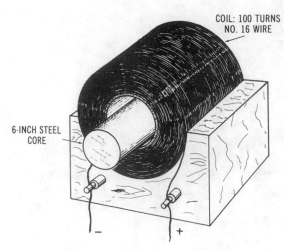

(a) Coil assembly with core inserted.

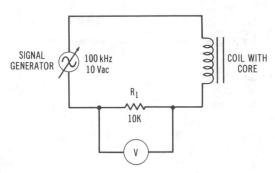

(b) Complete displacement circuit.

Fig. 4-3C. Inductive displacement transducer circuit.

Minimum R_1 voltage = _____ V

Maximum R_1 voltage = _____ V

4. Explain why the movement of the core of the coil affects the voltage across R_1.

_____

_____

_____

_____

_____

5. Construct the circuit in Fig. 4-3D.
6. Insert the movable core into L_2 and adjust its position, as well as the resistance of R_2, until the null state is achieved. (NOTE: You may find it necessary to insert a movable core into L_1 to achieve the null state.)
7. Slowly move the core of L_2 from null to the left and to the right. Describe how this affects the null indicator (ac ampere meter).

_____

_____

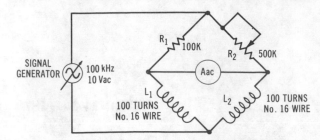

Fig. 4-3D. Inductive displacement transducer bridge circuit.

8. Return L_2 core to the null position. Assuming this core was mechanically coupled or linked to an object whose displacement was being monitored, complete Table 4-3A.
9. How does the current caused by the movement of the core to the left compare with the current caused when the core is moved to the right?

_____

_____

10. How could the circuit illustrated in Fig. 4-3D be made more sensitive?

_____

_____

_____

_____

Table 4-3A. Simulation of Inductive Displacement Transducer

Displacement Direction from Null	Distance of Displacement		AC Current Caused by Displacement
	Inches	Centimeters	
Left	1/16	0.159	
Left	1/8	0.318	
Left	3/16	0.477	
Left	1/4	0.635	
Left	5/16	0.793	
Left	3/8	0.953	
Return the core to the null position			
Right	1/16	0.159	
Right	1/8	0.318	
Right	3/16	0.477	
Right	1/4	0.635	
Right	5/16	0.793	
Right	3/8	0.953	

Analysis

1. What are some physical factors that determine the inductance of a coil?

   _____

   _____

   _____

2. How can the inductance of a coil be most easily altered?

   _____

   _____

   _____

3. Inductors are classified as air-core or iron-core. Which of these two types of inductors exhibits the greater inductance?

   _____

4. What is the relation between a core of an inductor and X_L?

   _____

   _____

   _____

5. How could an inductive displacement transducer be used with direct current?

   _____

   _____

Magnetic Displacement Transducers

Introduction

The most common magnetic displacement transducer is the linear variable differential transformer (lvdt), which is illustrated in Fig. 4-4A.

The lvdt employs a primary coil, a movable core, and two secondary coils connected in series. The movable core serves as an inductive or magnetic link between the primary and secondary coils. The amount of voltage induced from the primary coil to the secondary coils depends upon the position of the movable core. If the core is moved nearer secondary coil No. 1 than secondary coil No. 2, more induced voltage will appear across coil 1 than coil 2. If the movable core is moved nearer coil 2 than coil 1, the opposite will be true. Since the secondary coils are connected in series to oppose, their voltages are 180° out of phase. Thus, if the movable core is positioned on center relative to coil 1 and coil 2, an equal amount of voltage would be induced across each. These voltages would be 180° out of phase and would cancel. Therefore, when the movable core is centrally located, the voltage output of the lvdt is zero. The position of the movable core that results in a zero output of the lvdt is known as its *null position*.

The movable core of the lvdt is usually nulled and mechanically linked to the object whose displacement is to be measured. Any slight displacement of the tested object would cause a voltage to appear at the output of the lvdt. The magnitude of the voltage output would be indicative of the amount of displacement that had taken place.

Generally the sensitivity of an lvdt is expressed as voltage output per fraction of an inch displacement.

Objective

In this activity you will construct an lvdt and observe its operation.

Equipment

Digital vom
Dual-channel oscilloscope
6-Vac power supply
Air-core coils, 100 turns, A.S. No. 16 wire (2)
Air-core coil, 200 turns, A.S. No. 24 wire
6-inch steel core, ¾-inch diameter
Connecting wires

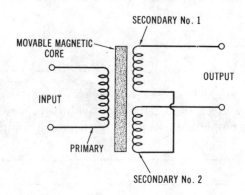

Fig. 4-4A. Linear variable-differential transformer.

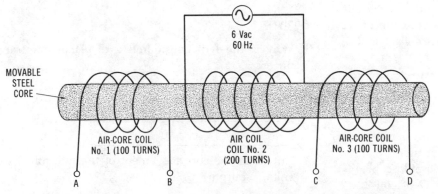

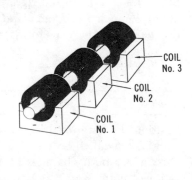

Fig. 4-4B. Construction of lvdt circuit.

Procedure

1. Use the three air-core coils, arranged as shown to construct the circuit in Fig. 4-4B. (NOTE: These coils should be placed as near each other as possible.)
2. Connect one channel of the oscilloscope to points *A* and *B* and the remaining oscilloscope channel to points *C* and *D*. (NOTE: The vertical attenuation adjustments for each scope channel should be the same.)
3. Adjust the horizontal sweep control until three or four sine waves appear on each channel of the scope. (NOTE: The waveforms appearing on channel 1 should be 180° out of phase with those appearing on channel 2. If the waveforms are not out of phase, reverse coil 1.)
4. Observing the proper phase relationship draw, in the space below, the waveforms that appear across coils 1 and 3.

Waveform across coil 1

Waveform across coil 3

5. Slowly slide the steel core from the left of center to the right about ½ inch (1.27 cm) and describe how this affects the waveforms across coils 1 and 3.

_____

_____

_____

6. Slide the core from the right of center to the left about ½ inch (1.27 cm) and describe the effect this has upon the waveforms across coils 1 and 3.

_____

_____

_____

7. Position the core until the waveforms across coils 1 and 3 appear to be equal in amplitude.
8. Disconnect both channels of the oscilloscope.
9. Connect point *B* of coil 1 to point *D* of coil 3.
10. Connect one channel of the oscilloscope to points

A and *C* and record the waveforms in the space below.

11. Slide the core about ½ inch (1.27 cm) from the center to the right and record the waveform appearing on the oscilloscope in the following space.

12. Slide the core about ½ inch (1.27 cm) from the center to the left and record in the space below the waveform appearing on the oscilloscope.

13. Slide the core to its null position and record in the space below the waveform as displayed by the scope.

14. How do the phases of the waveforms in Steps 11 and 12 compare?

_____

_____

_____

15. How does sliding the core within the coil affect the amplitude and phase of the voltage output?

_____

_____

_____

Analysis

1. Why is the lvdt considered a displacement transducer?

_____

_____

_____

2. How does the core movement of the lvdt affect the voltage output?

_____

_____

_____

3. How is the sensitivity of an lvdt expressed?

_____

_____

4. What is the null position of the core?

_____

_____

5. Why is the voltage output of the lvdt at zero when the core is in the null position?

_____

_____

_____

_____

LVDT Detector

Introduction

Frequently it is not enough to measure only the amount of the displacement of an object. Often the direction of displacement, as well as the amount of displacement, must be indicated. When this becomes the case, a detector circuit similar to that shown in Fig. 4-5A can be used in conjunction with the output of the lvdt.

In this detector circuit the diodes conduct equally when the movable core is in the null position. This results in an equal and opposite voltage drop across R_1 and R_2, causing the readout device (voltmeter) to indicate zero displacement. When displacement occurs, one diode will conduct more than the other, causing an upset in the balanced voltage across R_1 and R_2. The result is an output voltage, as measured by the voltmeter, whose polarity indicates the direction of displacement and whose magnitude indicates the amount of displacement.

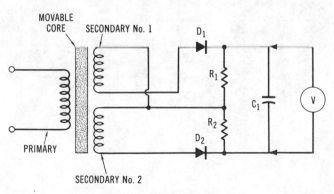

Fig. 4-5A. Lvdt detector circuit.

Objective

In this activity you will examine a detector circuit used with the lvdt constructed in the previous activity.

Equipment

Digital vom
Air-core coils, 100 turns, A.S. No. 16 wire (2)
Air-core coil, 200 turns, A.S. No. 24 wire

6-inch (15-cm) steel core, ¾-inch (1.91-cm) diameter
6-Vac power supply
1N4004 diodes (2)
Resistors: 1 kΩ (2)
Capacitor: 1 μF, 25 Vdcw
Connecting wires
Spst switch

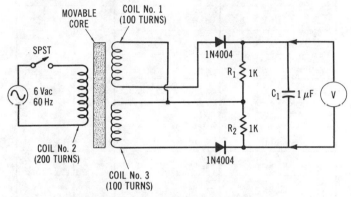

Fig. 4-5B. Test lvdt detector circuit.

Procedure

1. Construct the circuit illustrated in Fig. 4-5B.
2. Close the spst switch and slide the movable core until the voltmeter indicates null or zero.
3. Moving the movable core in the direction and by the amount indicated on Table 4-5A, complete the table by recording the voltage output of the detector circuit along with the proper voltage polarity.
4. How do the voltages generated by moving the core to the left compare with the voltages generated when the core is moved to the right?

_____

_____

5. Ideally, what is the minimum displacement of the movable core that would result in a voltage output?

_____

_____

_____

Analysis

1. What is meant by coils being connected in series to oppose?

Table 4-5A. Output Polarity and Magnitude of LVDT Test Circuit

Direction of Movement from Null Position	Amount of Movement		Voltage Output	Polarity (+ or −)
	Inches	Centimeters		
Left	1/16	0.159		
Left	1/8	0.318		
Left	3/16	0.477		
Left	1/4	0.635		
Left	5/16	0.793		
Left	3/8	0.953		
Return to Null Position				
Right	1/16	0.159		
Right	1/8	0.318		
Right	3/16	0.477		
Right	1/4	0.635		
Right	5/16	0.793		
Right	3/8	0.953		

_____

_____

2. Explain how the detector circuit used in conjunction with the lvdt enables the direction of displacement to be measured.

_____

_____

_____

_____

3. When is it necessary to measure direction, as well as amount of displacement?

_____

_____

4. How would the null position of the core be affected, if R_1 of the detector circuit were a value different from R_2?

_____

_____

_____

_____

Strain Gage Characteristics

Introduction

The strain gage is one of the most important transducers commonly used in the area of instrumentation. The strain gage along with appropriate electrical circuitry can be used in measurements of liquid and gas pressure, strain displacement, force, or weight.

The most common strain gage in use today is that constructed from fine wire or foil, and mounted on a flexible backing. Fig. 4-6A shows the physical arrangement of a strain gage.

Strain gages may also be constructed of a special type of flexible n-type semiconductor material. These are generally more sensitive than the wire type strain gages and are also more expensive.

The strain gage converts strain to resistance change. You will recall that the resistance of any conductor depends upon three factors:

1. Length (the longer the conductor, the greater its resistance).

FABT

FAB

FAES

FAED
FABD

FAE
FAB
FAP

FABX

FAER
FABR

Courtesy BLH Electronics, Inc.

Strain gages.

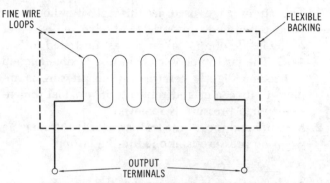

Fig. 4-6A. Physical structure of strain gage.

2. Cross-sectional area (the larger the cross-sectional area, the less its resistance)
3. Temperature (the higher the temperature of a conductor with a positive temperature coefficient, the greater its resistance).

If a certain force is applied to the strain gage illustrated in Fig. 4-6B in its active direction, its conductor would become shortened and increase in cross-sectional area. This would result in the resistance of the strain gage being decreased proportionally to the

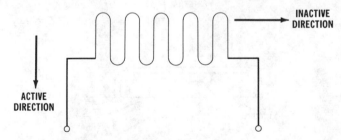

Fig. 4-6B. Active and inactive directions of a strain gage.

force that caused the elongation. Thus the resistance of a strain gage is indicative of the pressure or force applied to the surface to which the strain gage is bonded.

Objective

In this activity you will observe and examine the characteristics of a strain gage.

Equipment

Digital vom
Strain gage: 120 Ω, GF = 2.1
Metal jar top
Adhesive
Connecting wires

Procedure

1. The strain gage used in this activity should be bonded, with an appropriate adhesive, to the center of the inside of a metal jar top as illustrated in Fig. 4-6C. The jar top chosen should be flexible enough to become slightly deformed when pressure is applied to its center and return to its original dimensions when pressure is removed.

2. Measure and record the resistance of the strain gage when no pressure is applied to the jar top.

$R = \underline{\hspace{2cm}} \Omega$

3. Apply pressure to the center of the jar top as illustrated in Fig. 4-6D. Record the resistance of the strain gage caused by this pressure.

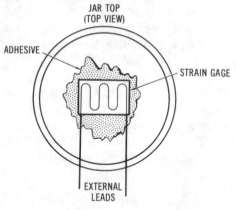

JAR TOP
(TOP VIEW)

ADHESIVE

STRAIN GAGE

EXTERNAL
LEADS

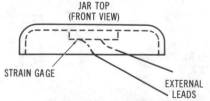

JAR TOP
(FRONT VIEW)

STRAIN GAGE

EXTERNAL
LEADS

Fig. 4-6C. Connecting strain gage to jar top.

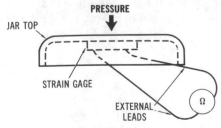

PRESSURE

JAR TOP

STRAIN GAGE

EXTERNAL
LEADS

Ω

Fig. 4-6D. Applying pressure to jar top.

4. Increase the pressure to the jar top, as illustrated in Fig. 4-6D, until the strain gage exhibits its maximum resistance. (NOTE: Do not apply enough pressure to permanently deform the jar top.) Record this resistance in the space below.

$R = \underline{\hspace{2cm}} \Omega$

5. How do the resistances recorded in Steps 2, 3, and 4 compare?

_____

_____

_____

6. How does pressure applied to the surface to which the strain gage is bonded affect strain gage resistance?

_____

_____

_____

7. Is the change in resistance of a strain gage, which is due to pressure, small or large?

_____

_____

_____

Analysis

1. For what are strain gages used?

_____

_____

_____

2. Pressure or force may be applied to a strain gage in the active or inactive direction. Describe the effect upon strain gage resistance when a force is applied to it in the inactive direction.

_____

_____

_____

3. What three factors determine the resistance of any metallic conductor?

_____

_____

_____

4. Describe how force causes a change in strain gage resistance.

_____   _____

_____   _____

_____   _____

_____   _____

Strain Gage Application

Introduction

The resistance change of a strain gage is caused by a force applied to the surface to which the strain gage is bonded. This force actually causes the conductor associated with the strain gage to change its length. This results in a new length of strain gage conductor that can be computed by adding the change in length to the length of the conductor when no force is applied (original length plus change in length equals the new length). Since strain is a unitless quantity, it is commonly expressed as a ratio of change in length to length and it is expressed in micro (μ) units:

$$\text{strain} = \frac{\Delta L}{L}$$

where

ΔL is a change in the length of the conductor caused by a force,
L is the original length of the conductor before being changed by a force.

The length change caused by a force applied to a strain gage brings about a change in its resistance. The resistance of the strain gage with force applied can be computed by adding the change in its resistance with force applied to its resistance before force is applied (resistance before force is applied plus resistance change caused by force equals the resistance of the strain gage with force applied). The *resistance coefficient* of a strain gage is expressed as a ratio of change in resistance to resistance:

$$\text{resistance coefficient} = \frac{\Delta R}{R}$$

where

ΔR is a change in resistance caused by a change in length of the strain gage conductor,
R is the resistance of the strain gage when no force is applied.

The relation between resistance coefficient ($\Delta R/R$) and strain ($\Delta L/L$) is usually expressed as a ratio of those quantities and is called the *gage factor*:

$$\text{gage factor (GF)} = \frac{\Delta R/R}{\Delta L/L} \quad \text{or} \quad \frac{\Delta R/R}{\text{strain}}$$

where the right-hand quantities are as defined previously.

The gage factor is indicative of the relation existing between resistance and length. This information is provided, along with strain gage resistance in ohms, by most manufacturers. The gage factor of most metallic strain gages is around 2, while typical resistances range from 60 to 1000 ohms.

By knowing the gage factor and resistance of a strain gage, resistance change can be computed by using the appropriate equation:

$$\Delta R = (\text{GF})(\text{strain})(R)$$

If the resistance coefficient and gage factor are known, then

$$\text{strain} = \frac{\Delta R/R}{\text{GF}}$$

may be calculated.

Strain gages are most commonly used in pressure or weight measuring applications. The strain gage can be connected as one arm of a four-arm bridge circuit whose null indicator is calibrated in pounds per square inch, pounds, grams, etc., and is read directly.

Objective

In this activity you will construct two circuits utilizing a strain gage. You will observe the action of these circuits when pressure is applied to the strain gage.

Equipment

Digital vom
Strain gage: 120 Ω, GF = 2.1
Resistors: 100 Ω (2), 680 Ω, 6.8 kΩ
Potentiometer: 200 Ω
Dc power supplies (2)
Transistors: GE-FET-1, 2N2405

Metal jar top
Connecting wires

Procedure

1. Construct the circuit in Fig. 4-7A. (NOTE: The strain gage and jar top assembly used in the previous activity should be used here.)
2. Adjust R_3 until the null condition is achieved.
3. Apply pressure to the center of the jar top and describe how this affects the reading of the vom.

_____

_____

_____

4. Apply pressure to the center of the jar top and record the current as indicated by the digital vom.

$I =$ _____ mA

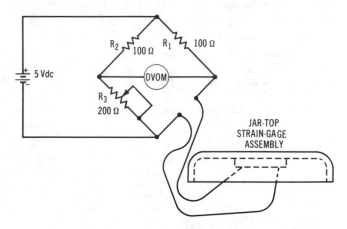

Fig. 4-7A. Strain-gage bridge circuit.

5. Increase the pressure to the jar top and describe how this affects the reading of the vom.

_____

_____

_____

6. Alter the circuit illustrated in Fig. 4-7A to that in Fig. 4-7B.
7. Record the collector current of Q_2 when the bridge circuit is nulled and no pressure is applied to the jar-top assembly.

$I_C =$ _____ mA

8. Apply pressure to the center of the jar-top assembly and record the resulting collector current of Q_2.

$I_C =$ _____ mA

9. Increase and decrease the pressure being applied to the jar top and record the range of resulting collector currents of Q_2.

I_C is from _____ mA to _____ mA

10. Describe the effect upon the collector current when pressure is applied to the strain-gage jar-top assembly.

_____

_____

_____

11. If the meter connected to measure collector current were properly calibrated, how could this circuit be used to measure pressure?

_____

_____

_____

Analysis

1. List some of the applications of the strain gage.

_____

_____

_____

2. The strain gage used in this activity has a resistance of 120 Ω and a gage factor of 2.1. What strain will bring about a change in its resistance of 1 Ω?

_____

3. What change in resistance will result when a strain of 1500 μm/m is applied to the strain gage?

_____

4. What is the resistance of a 500-Ω strain gage with a gage factor of 2 when a strain of 2000 μm/m is applied?

_____

5. Write an analysis of the operation of the circuit illustrated in Fig. 4-7B of this activity.

_____

_____

_____

_____

_____

_____

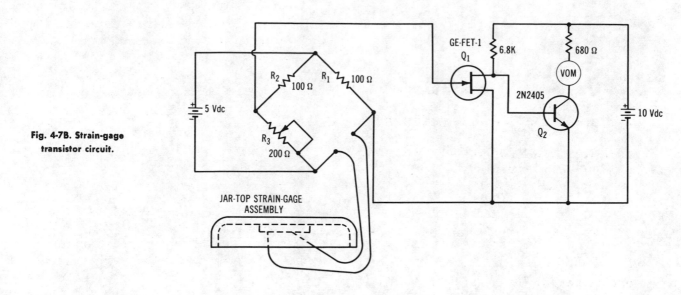

Fig. 4-7B. Strain-gage transistor circuit.

Strain-Gage Temperature Compensation

Introduction

The resistance of any metallic conductor is affected by the temperature of its environment. As the temperature of a metallic conductor increases, the resistance of the conductor likewise increases.

Since the resistance change caused by force upon a strain gage is small, any change in its environmental temperature can result in an additional resistance change unrelated to the force. This usually causes a measurement error in pressure, weight, etc.

Strain gages are used in industrial environments whose temperatures range widely. In order to maintain measurement accuracy the resistance change in strain gages caused by operating temperature variations must be corrected. This is normally accomplished by using an active strain gage along with an inactive strain gage that acts to compensate for temperature variations.

Objective

In this activity you will examine a circuit that employs a temperature-compensating strain gage.

Equipment

Digital vom
Strain gages: 120 Ω, GF = 2.1 (2)
Adhesive
Resistor: 100 Ω
Potentiometer: 200 Ω
Ac-dc power supply
60-W lamp with socket
Metal jar top
Connecting wires

Procedure

1. Using an appropriate adhesive, glue a second strain gage of identical characteristics to the edge of the jar top used in Activities 4-6 and 4-7. See Fig. 4-8A.
2. The original strain gage (inside the jar top) will be considered the active strain gage while the one

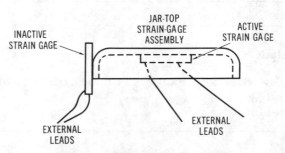

Fig. 4-8A. Attaching inactive strain gage.

glued to the jar-top edge will be considered the inactive gage.
3. Construct the circuit illustrated in Fig. 4-8B.
4. Adjust R_1 until the null state is indicated by the digital vom.
5. Apply pressure to the center of the jar top and record the resulting current.

$I = \underline{\hspace{1cm}}$ mA

6. Pressure applied to the active strain gage results in the unbalance of the bridge circuit. The inactive gage is not affected by the applied pressure; thus its resistance remains independent of pressure. Since both the active and inactive strain gages are mounted near each other, their resistance would be equally affected by a temperature change. This would cause the null state of the bridge to be maintained even though the temperature of the strain gages might change. Any unbalance of the bridge

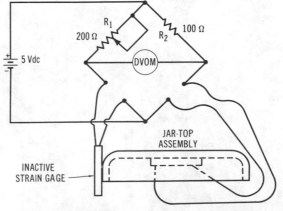

Fig. 4-8B. Temperature-compensating strain-gage circuit.

185

would be due to pressure applied to the active strain gage.

7. Position the 60-watt lamp near the jar top assembly. Connect 120 Vac to the lamp and allow it to warm the jar-top assembly several degrees.

8. How did the increased temperature caused by the lamp affect the null state of the bridge?

_____

_____

_____

Analysis

1. How does temperature affect the resistance of a metallic strain gage?

_____

_____

2. Explain how the inactive strain gage used in this activity compensated for any resistance variation caused by the temperature of the active strain gage.

_____

_____

_____

_____

3. Why is temperature compensation necessary for strain gages?

_____

_____

Measurement of Physical Quantities

Unit 5 presents some practical applications of instrumentation circuitry. The measurement of physical quantities usually is dependent upon some type of transducer that converts the physical quantity into an electrical quantity. The electrical quantity is then calibrated to correspond to a meter reading of some type. When the meter scale is calibrated properly, the physical characteristic can be measured directly by the meter. Transducers are used with instruments to measure almost any physical quantity.

In the following activities you will investigate the measurement of some specific physical quantities. The measurement of physical quantities is becoming more important than ever for instrumentation systems due to the increased precision of process control and automation. Some of the common types of measurement involving physical quantities are studied in this unit. These physical measurements include humidity, pressure, flow of gas and liquid, bulk and liquid level, sound, pH, and time.

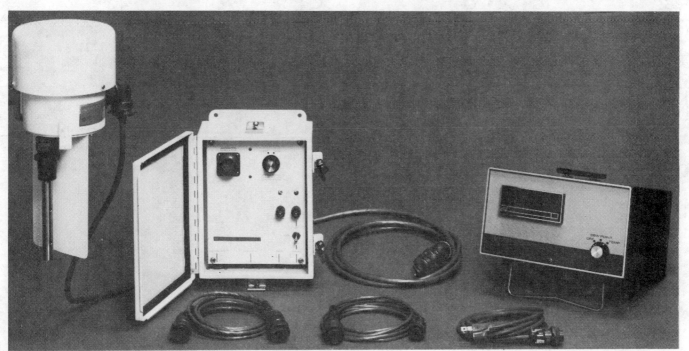

Courtesy General Eastern Instruments Corp.

Meteorological dew point hygrometer system.

Humidity Measurements

Introduction

"Humidity" is a term used to describe the amount of moisture or water in air or other gases. The measurement and control of humidity is vitally important in a number of industrial settings. Humidity measurement is especially important in the paper, tobacco, cement, and textile industries.

The moisture content of air may be expressed as the absolute or relative humidity. Relative humidity is the percentage of moisture in the air as compared to the moisture of saturated air at the same temperature. As the temperature of air increases, likewise will the ability of the air to hold moisture increase. A knowledge of the temperature of the air is essential when measuring relative humidity. A relative humidity of 100 percent means that air holds all of the moisture it can at its specified temperature. Absolute humidity is expressed as grains of moisture per pound (453.6 grams) of air (one grain equals 1.42857×10^{-4} pound or 0.065 gram) and represents the amount of moisture that is, in fact, present in any volume of air.

Since humidity depends largely upon air temperature, it is essential that temperature be measured accurately. There are three classifications of temperature which relate to humidity measurements. These are the dry bulb, wet bulb, and dew point temperatures. The *dry bulb* temperature of air is simply its temperature in degrees Fahrenheit or Celsius as measured with a mercury-in-glass thermometer. The *wet bulb* temperature is the temperature of the air at saturation. The *dew point* temperature represents the temperature at which air will begin depositing dew. If the dew point of air is below 32°F, it is generally referred to as the *frost point*.

Basic and essential to all humidity measurements is a psychrometric chart or graph similar to the one illustrated in Fig. 5-1A. The psychrometric chart represents the relationship that exists among wet and dry bulb temperatures, relative humidity, and absolute humidity. To use this figure one must first determine the dry and wet bulb temperatures. These temperatures are then projected along their respective lines until they intersect. This intersection point represents

Single-stage thermoelectric condensation hygrometer.

Courtesy General Eastern Instruments Corp.

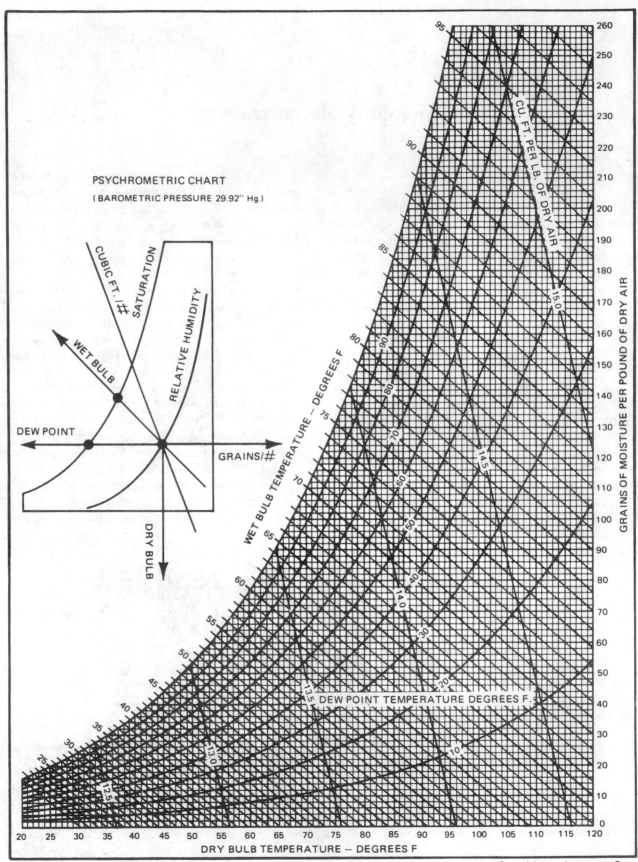

Fig. 5-1A. Psychrometric chart (barometric pressure 29.92 inches Hg).

Dry-bulb/wet-bulb psychrometer system with temprature indicator.

Courtesy General Eastern Instruments Corp.

the relative humidity. Projecting horizontally from this point of intersection determines the absolute humidity. It can be seen from the chart that when the dry and wet bulb temperatures are equal, the air is saturated and the relative humidity is 100 percent.

Objective

In this activity you will learn to use a simple sling psychrometer to measure both dry bulb and wet bulb temperatures. With these temperatures and a psychrometric chart both the relative and absolute humidity can be determined.

Equipment

Sling psychrometer (20°F to 120°F or −6.6°C to 48.8°C)

Psychrometric chart (20°F to 120°F or −6.6°C to 48.8°C)

Procedure

1. Examine the sling psychrometer provided for this activity. List the minimum and maximum dry and wet bulb temperatures that can be measured with this instrument.

Dry bulb temperatures: _____ min;

_____ max

Wet bulb temperatures: _____ min;

_____ max

2. Fill the reservoir with clean water, at room temperature if possible, until the wick on the wet bulb thermometer is completely saturated.

3. Spin the thermometer part of the sling psychrometer at a rate of about two revolutions per second for 15 to 20 seconds. Quickly record both the wet and dry bulb temperatures as exhibited by the appropriate thermometer.

Dry bulb temperature = _____ °F

Wet bulb temperature = _____ °F

4. Using the data gathered in Step 3 along with the psychrometric chart, determine the relative and absolute humidity.

Relative humidity = _____ percent

Absolute humidity = _____ grains per pound

5. Using the procedure as previously outlined, complete Table 5-1 by gathering data relative to humidity from five different environments. (One location should be out-of-doors if possible.)

Sling psychrometer.

Courtesy J. R. Douglas Co.

Table 5-1. Humidity Measurements and Data

Environment	Wet Bulb Temperature (°F)	Dry Bulb Temperature (°F)	Relative Humidity (%)	Absolute Humidity (grains per lb)

Analysis

1. Define the following terms:
   (a) Humidity:

   _____

   _____

   _____

   (b) Relative humidity:

   _____

   _____

   _____

   (c) Absolute humidity:

   _____

   _____

   _____

   (d) Wet bulb temperature:

   _____

   _____

   _____

   (e) Dew point temperature:

   _____

   _____

   _____

   (f) Dry bulb temperature:

   _____

   _____

   _____

(g) Grains per pound of air:

_____

_____

_____

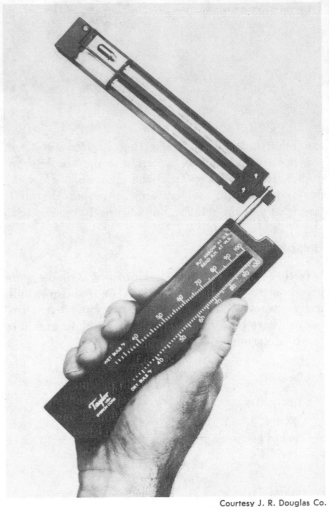

Courtesy J. R. Douglas Co.

Spinning the sling psychrometer.

2. What is meant by air saturation?

   _____

   _____

   _____

3. What condition causes the wet bulb and dry bulb temperatures to be equal?

   _____

   _____

   _____

4. Why are humidity measurements important?

Measuring Pressure With a Manometer

Introduction

The measurement of pressure can be accomplished by several different methods using a variety of transducers. Applications of pressure measurement are very numerous. Pressure is defined as a force per unit area.

One way to measure pressure is with a U-tube manometer, illustrated in Fig. 5-2A. The pressure created by a fluid forced into the manometer enclosure is used to balance with an unknown pressure. If only air pressure is applied to the manometer, the two columns of liquid will be at the same height above an established reference (points A and B). However, if a fluid is directed into the left column of the instrument, it exerts a certain pressure (P_1). The level of P_2 will then rise until an equilibrium condition is established (points C and D). The difference between points B and D is then a measure of the pressure directed into the left column. If a heavy liquid such as mercury is used, very high pressures can be measured. There are several variations of the basic U-tube manometer for measuring different types and ranges of pressure.

Another method of measuring pressure relies on a capacitive transducer. An increase in pressure of a fluid within an enclosure will cause the movable plate of the capacitor to come closer to the stationary plate. The capacitance will then increase. The scale

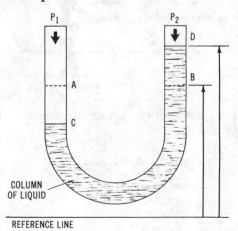

Fig. 5-2A. U-tube manometer.

of the indicator must be calibrated to measure the change in capacitance in terms of the fluid pressure.

A strain-gage transducer can also be used to measure pressure, as illustrated in Fig. 5-2B. When increased pressure is exerted on the flexible steel from the pressure source, the strain-gage wire will change dimensions. The change in resistance due to the

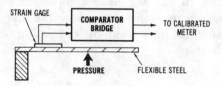

(a) A strain gage.

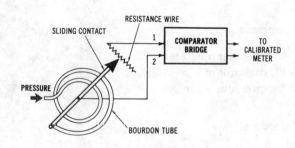

(b) A Bourdon tube.

Fig. 5-2B. Methods of measuring pressure.

change of dimension of the strain gage is sensed by a comparative bridge network. The output of the bridge supplies current to cause deflection on a meter calibrated to measure changes of pressure exerted on the strain gage.

A Bourdon-tube spiral spring, such as the one illustrated in Fig. 5-2B, can also be used to measure pressure. The Bourdon tube is a highly elastic metallic element. Its elastic deformation is proportional to the applied pressure. In the arrangement shown, as the pressure increases, the Bourdon-tube spring tends to straighten out, causing movement of a sliding contact connected to a fixed amount of resistance wire. The movement of the sliding contact will affect the resistance between points 1 and 2. These two points are

PLASTIC TUBING

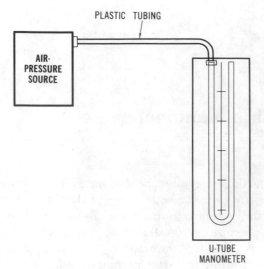

Fig. 5-2C. U-tube manometer setup.

connected to one leg of a comparator bridge. The difference in resistance caused by the increased pressure can then be sensed by the calibrated scale of the indicator.

Objective

In this activity you will use a U-tube manometer to measure pressure.

Equipment

U-tube manometer
Plastic tubing
Air-pressure source

Procedure

1. Set up the manometer circuit as shown in Fig. 5-2C.
2. Record the water level on each side of the manometer with no external pressure applied.

   Right side = _____

   Left side = _____
3. Increase the external pressure slightly. What is the effect on the water level on each side of the manometer?

   _____

   _____

   _____
4. Increase the external pressure until there is a distinct difference between the water levels. Measure and record the difference in these two levels.

   Difference = _____

Courtesy Dwyer Instruments, Inc.

Manometer.

5. Decrease the external pressure and note the effect on the water level.
6. This concludes the activity.

Analysis

1. Define pressure.

   _____

   _____

   _____

2. How can atmospheric pressure be measured?

   _____

   _____

   _____

3. Discuss the use of the U-tube manometer for measuring pressure.

   _____

   _____

   _____

   _____

4. What is Pascal's law?

_____

_____

_____

_____

_____

5. What are some other methods of measuring pressure?

_____

_____

_____

_____

Gas Flow Measurements

Introduction

The measurement of the flow of various types of gases is important in many of the manufacturing processes, as well as in situations where an artificial atmosphere or environment must be maintained. The flow rate of oxygen is of vital importance to patients in respirators and oxygen tents. The flow rate of inert gases and/or oxygen into ovens and furnaces where combustion takes place, in many instances, controls the internal temperature of that chamber. In order that the flow rate of gas be controlled, it must first be measured.

There are a number of widely used methods for measuring gas flow. One method involves the use of two thermistors connected in a bridge circuit like the one shown in Fig. 5-3A. In this circuit thermistor No. 1 is mounted in a closed container and thermistor No. 2 is mounted inside a small pipe or tube. There is no air movement around thermistor No. 1. When there is no air movement through the pipe, thus around thermistor No. 2, the bridge circuit is balanced. When air is forced to move through the pipe, thermistor No. 2 is cooled proportionally by the air movement, and the bridge circuit becomes unbalanced. The unbalanced state of the bridge is indicative of the rate of air flow through the pipe.

Another method of measuring gas flow involves the use of a simple flowmeter. Generally these devices are constructed with a gas inlet and gas outlet, a tapered chamber, and a float. As gas moves into the tapered measuring chamber through the inlet, the float is forced upward. The height of the float is indicative of the rate of gas flow. Usually there are calibrating marks along the tapered measuring chamber which allow for direct reading of gas flow in cubic feet per minute (cfm) or cubic feet per hour (cfh) or static cubic feet per hour (scfh).

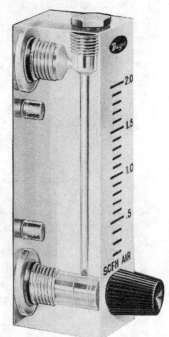

Air flowmeter.

Courtesy Dwyer Instruments, Inc.

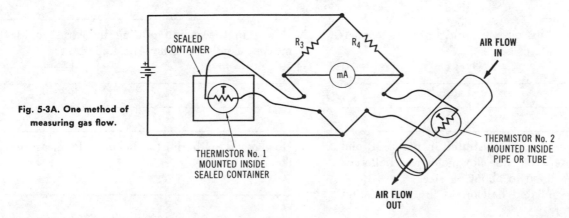

Fig. 5-3A. One method of measuring gas flow.

Objective

In this activity you will examine some of the basic principles of measuring gas flow.

Equipment

Digital vom
Tapered-tube flowmeter
Thermistors: Fenwal Electronics JA35J1 (2)
Resistor: 10 kΩ
Potentiometer: 25 kΩ
Dc power supply
Small balloon
Connecting wires

Procedure

1. Construct the circuit shown in Fig. 5-3B.
2. Adjust R_1 until the null state is reached.

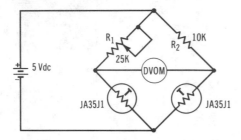

Fig. 5-3B. Use of thermistor bridge.

3. Shield R_3 and force air to move across R_4. Describe how this action affects the null state of the circuit.

_____

_____

_____

4. Increase the rate of air flow across R_4 while shielding R_3. Record the current measured by the meter.

   $I =$ _____ mA

5. Describe how this circuit could be used to measure the flow rate of gas.

_____

_____

_____

6. Disconnect the circuit illustrated in Fig. 5-3B and acquire the tapered tube flowmeter and balloon.
7. Inflate the balloon to about ⅛ size.
8. Connect the inflated balloon to the gas inlet of the flowmeter.

9. Allow the air from the balloon to discharge through the flowmeter and record the rate of air flow as defined by the position of the float during discharge.
   Air flow rate = _____ scfh
10. Inflate the balloon to ¼ size and repeat Steps 8 and 9.
    Air flow rate = _____ scfh
11. What is the relationship among the float, tapered tube, and rate of air flow?

_____

_____

_____

Courtesy Dwyer Instruments, Inc.

Three air flowmeters.

Analysis

1. Where in the medical field would the measurement of oxygen flow be important?

_____

_____

2. Where in the manufacturing industry would the measurement of gas flow be important?

_____

_____

3. How could a thermistor be used to measure gas flow?

_____

_____

4. Explain how the simple tapered tube flowmeter measures air flow.

_____

_____

_____

_____

5. What are some common units of measurement for gas flow?

_____

_____

_____

_____

Measuring Liquid Flow

Introduction

The measurement of the flow of liquids and gases is important for many applications. Many industrial processes rely upon specific flow rates for proper operation. As with most physical quantities, flow rates can be measured by utilizing various types of transducers. A common type of flowmeter is the magnetic flowmeter, illustrated in (a) of Fig. 5-4A. It is designed to measure the flow rate of electrically conductive fluids. The meter is placed into the fluid line so that all the fluid flowing in the line passes through it. Since the fluid line contains a conductive fluid, the fluid between the coil terminals produces a magnetic field which induces a voltage into the coil. The faster the rate of fluid flow, the higher in magnitude the voltage becomes. Thus the voltage developed in the coil is proportional to the fluid flow rate. The coil output is connected to a calibrated meter.

Another method of measuring flow rate utilizes a thermistor as the transducer. The thermistor, as shown in (b) of Fig. 5-4A, is placed within the fluid line so that the fluid passes by it. A dc potential is applied to the thermistor to cause it to be heated. If there is no fluid flow, the thermistor is cooled at a uniform rate by the fluid, causing a constant reading on the current meter connected in series with the thermistor and voltage source. As the flow rate increases, the thermistor will decrease in temperature in proportion to the rate of fluid flow. As the temperature of the thermistor decreases, its resistance increases. The reading on the current meter will then decrease due to the increased resistance. The meter scale is calibrated to measure flow rate. A commercial flowmeter is shown in Fig. 5-4B.

Courtesy Foxboro Co.

Fig. 5-4B. A commercial flowmeter.

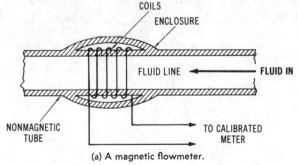

(a) A magnetic flowmeter.

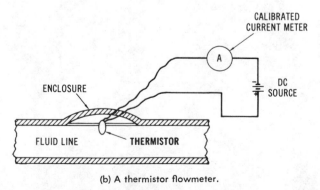

(b) A thermistor flowmeter.

Fig. 5-4A. Methods of measuring flow rate.

Objective

In this activity you will construct a circuit using a thermistor transducer which could be used to measure liquid flow.

Equipment

Thermistor: JA41J1 or equivalent
Dc power source
Electronic multifunction meter

Plastic container
Plastic tubing
Funnel
Resistor: 1 kΩ
Water
Connecting wires

Procedure

1. Construct the circuit and experimental setup for measuring liquid flow shown in Fig. 5-4C.
2. Make an opening in the plastic tubing large enough for the thermistor to press into. The tubing should be fastened securely to the plastic container. The funnel is used to direct water onto the thermistor.
3. Turn on the dc power source. Record the current reading of the "flow indicator."

$I =$ _____ mA

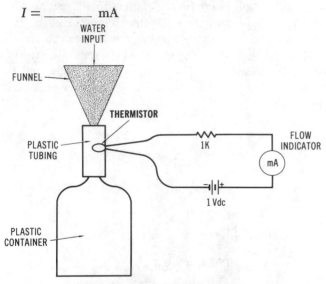

Fig. 5-4C. Setup for measuring liquid flow.

4. Pour a small amount of water through the funnel into the plastic container.
5. Record the amount of current through the "flow indicator."

$I =$ _____ mA

6. Pour more water through the funnel.
7. Record the new value of current.

$I =$ _____ mA

8. This concludes the activity.

Analysis

1. Compare the values of current recorded in Steps 3, 5, and 7. Why are they different?

_____

_____

_____

2. How could the sensitivity of the thermistor circuit be improved?

_____

_____

_____

3. What are some methods of measuring liquid flow?

_____

_____

_____

_____

_____

_____

Bulk Level Measurements

Introduction

Bulk level measurements deal with the measurements of the level of materials such as grain, cereal, tobacco, cement, gravel, and coffee, as these materials are placed in vats or hoppers during the manufacturing or preparation process. These measurements may be achieved mechanically or electrically and are vitally important in any industry involved with the processing and packaging of bulk materials or products.

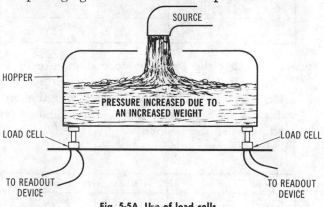

Fig. 5-5A. Use of load cells.

Load cells may be used to indicate the level of certain bulk materials in containers (see Fig. 5-5A). These devices contain strain gages. The resistance of the strain gage is altered due to the pressure that is caused by the weight of a material in a hopper or container.

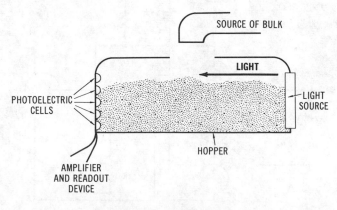

Fig. 5-5B. Use of photoelectric cells.

Courtesy BLH Electronics, Inc.

Load cells.

Bulk level may also be measured using light and appropriate sensors along with electrical circuitry. The electrical resistance of a group of series photoelectric cells, placed in a hopper, would be altered by the level of the bulk material in the hopper (Fig. 5-5B). This alteration in resistance would be indicative of the level of the material.

Objective

In this activity you will examine a photoelectric circuit that can be used to measure the bulk level of a material in a hopper or other large container.

Equipment

Digital vom
Photoconductive cells: GE-X6 (3)
Resistors: 100 kΩ, 1 kΩ
Potentiometer: 500 kΩ
Transistor: 2N2405
Ac-dc power supply
60-W lamp with socket

Connecting wires
Spst switch

Procedure

1. Construct the circuit in Fig. 5-5C.
2. Adjust R_2 to about midrange, place the 60-watt lamp about 6 inches (15 cm) from the series photoelectric cells and close the spst switch. Record the voltage as indicated by the voltmeter.

$V =$ _____ V

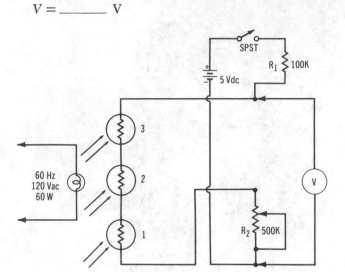

Fig. 5-5C. Bulk level measuring photoelectric circuit.

3. Pass your hand between the lamp and the photoconductive cells and describe the effect upon the voltage as indicated by the voltmeter.

_____

_____

_____

4. Using a piece of paper, prevent the light from falling upon the photoconductive cells as indicated below and record the voltage as measured by the voltmeter.

PC Cells in Darkness	Voltage
PC 1	
PC 1 and 2	
PC 1, 2, and 3	

5. Alter the circuit shown in Fig. 5-5C to that in Fig. 5-5D.
6. Adjust R_2, as well as the distance between the photoconductive cells and lamp, until Q_1 is at its

cutoff point but on the threshhold of conduction. This should cause the milliammeter to indicate zero current.

7. Place your hand between the lamp and the photoconductive cells and describe how this action affects the collector current of Q_1.

_____

_____

_____

8. Using a piece of paper, prevent the light from falling upon the photoconductive cells as indicated in the following. Record the collector current of Q_1 for each condition.

PC Cells in Darkness	Collector Current of Q_1
PC 1	
PC 1 and 2	
PC 1, 2, and 3	

9. Assume that the series photoconductive cells illustrated in Fig. 5-5D are positioned vertically in a large hopper with a light source. Using the data gathered in Step 8, indicate the milliammeter reading when the hopper is ⅓ filled, ⅔ filled, and filled with some grainlike material.

Hopper	Milliamperes
1/3 Filled	
2/3 Filled	
Filled	

10. How could the circuit illustrated in Fig. 5-5D be altered to provide information concerning when

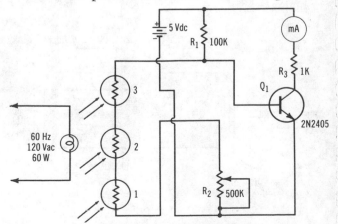

Fig. 5-5D. Transistor photoelectric circuit.

the hopper was ¼ filled, ½ filled, ¾ filled, and filled?

_____

_____

_____

Analysis

1. What is a load cell?

_____

_____

2. How could a load cell be used to measure bulk level?

_____

_____

3. What are some advantages and disadvantages of using the photoconductive cell circuit illustrated in Fig. 5-5C for bulk level measurements?

_____

_____

_____

_____

_____

4. Compare the circuits in Figs. 5-5C and 5-5D as bulk level measuring circuits.

_____

_____

_____

_____

_____

Liquid Level Control

Introduction

Many industrial processes rely upon liquid level measurement. Variables, such as fuel supply for example, are sometimes monitored continuously. There are several techniques which may be used to measure liquid level.

The measurement of liquid level is easy to accomplish by using transducers. Level changes result in the displacements of the top surface of the liquid. Many types of transducers may be used to measure liquid level. Resistive transducers could be used to measure the level of a conductive solution. A capacitive transducer could be used which employs a movable plate whose position is determined by the level of the liquid. Photoelectric methods, radioactive methods, and ultrasonic methods may also be employed.

A simple type of level controller is the ball float system. This system uses a ball float to operate a lever. The lever is connected to a valve which regulates liquid flow rate. Chemical industries commonly use differential pressure controllers for controlling the level of volatile liquids. The liquid pressure is proportional to its level in the enclosed container.

Level control may be accomplished by placing a light source at the same height above a conveyor line as the desired level of fill of a container. In the illustration of Fig. 5-6A containers on a conveyor line are positioned under a liquid dispenser. When the container is in position the actuator causes the dispenser to allow liquid to pass into the container. When the liquid reaches the level of the light source, the light beam is interrupted. With no light striking its surface, the detector will cause a relay to activate. The activated relay will, in turn, cause the actuator on the dispenser to close. When another container is in position, the actuator will open once more. This liquid level control system ensures a uniform level of liquid in each container.

Objective

In this activity you will construct and test a photoelectric circuit which could be used to measure liquid level.

Equipment

Variable dc power supply
Light-dependent resistor (GE-X6 or equivalent)
Electronic multifunction meter
Container (glass or plastic)
12-V, 1250-Ω relay (Guardian 1335-2C-120D or equivalent)
60-W lamp with holder
7-W lamp with holder
Spst switch
120-Vac power source

Procedure

1. Construct the photoelectric liquid level control of Fig. 5-6B.
2. Plug the light source into a 120-Vac power outlet and close the circuit switch. Position the light

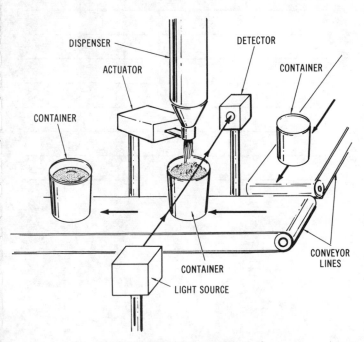

Fig. 5-6A. Liquid level control method.

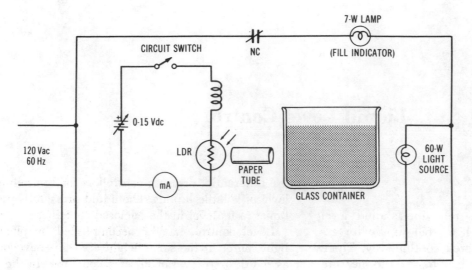

Fig. 5-6B. Photoelectric liquid level control.

source near the bottom of the glass container before it is filled with water. Slide the paper tube around the light-dependent resistor (ldr) and place the open end against the container.

3. If the circuit is operating properly, the fill indicator lamp will light when the relay is actuated.

4. Adjust the dc source to alter the sensitivity of the circuit. Do not exceed 15 Vdc. The circuit should be able to detect a pencil passing in front of the paper tube.

5. Carefully fill the container with water until the indicator turns off. Avoid pouring water directly in front of the tube window area. You may need to try several trial runs to get the sensitivity to a level where the circuit will respond properly.

6. After the sensitivity has been adjusted, drain or siphon water from the container until the fill indicator is actuated again.

7. Test the liquid level control circuit two or three times to assure yourself that it operates properly.

8. This concludes the activity.

Analysis

1. Discuss the operation of the circuit used in this activity.

2. What are some other types of circuits which could be used for liquid level control?

Telemetry Measurements

Introduction

When a quantity being measured is indicated at a location some distance from its transducer or sensing element, the measurement process is referred to as *telemetering*. Many types of indicating systems fit this definition; telemetering systems, however, are usually for long-distance measurement or for centralized measurement. For instance, many industries group their indicating systems together to facilitate process control. Another example of telemetering is the centralized monitoring of electrical power by utility companies on a regional basis. These systems are similar to other measuring systems except that a transmitter/receiver communication process is usually involved.

Many types of electrical and physical quantities can be monitored by using telemetering systems. The most common transmission media for telemetering systems are: (1) wire, such as telephone or telegraph lines; (2) superimposed signals, which are 30- to 200-kHz signals carried on power distribution lines; and (3) radio-frequency signals, mainly from am, fm, and phase-modulation transmitters. A type of telemetering system is shown in Fig. 5-7A. In the system shown, a dc voltage from the transducer could be used to modulate an am or fm transmitter. The rf signal is then received and converted back into a dc voltage to activate some end device. The end device, which may be located a considerable distance from the transducer, could be a chart recorder, a hand-

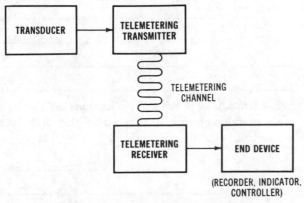

TELEMETERING CHANNEL

(RECORDER, INDICATOR, CONTROLLER)

Fig. 5-7A. Basic telemetering system.

deflection indicator, or possibly a process controller. Digital telemetering is also used since binary signals are well suited for data transmission. In this system the transducer output is converted to binary code for transmission.

Objective

In this activity you will construct a simple circuit that illustrates a telemetering system. Telemetering takes place when a change in one electrical or physical variable is transmitted by some means over a relatively long distance from where the change occurs. The amount of change is noted on the receiving end of the system by observation of some type of recording or monitoring device.

This activity uses two 555 integrated-circuit (IC) timers. They are used to sense a change in temperature of a thermistor, which causes a frequency change in the circuit. This frequency change can be noted on an am radio receiver located a long distance away.

The 555 IC can produce time delays from a few microseconds to hours. It can run as an oscillator at frequencies as low as one pulse per hour or as high as one megahertz. The internal structure of a 555 is composed of many transistors and other components, arranged to make up the following circuits: one bistable multivibrator (flip-flop), two comparators, and an output stage. The oscillation capability of the 555 is used in this activity. Its frequency is changed by an external *RC* network in which the resistance is a temperature sensitive thermistor. Thus changes in temperature can be transmitted over a long distance to an am radio receiver and noted as changes in frequency.

Equipment

Integrated circuits: 555 IC timers with sockets (2)
Oscilloscope
Dc power sources (2)
Capacitors: 24 pF, 0.01 μF, 1.5 μF
Resistors: 650 Ω, 1000 Ω (2)
Thermistor: 1000 Ω (KA31L1 or equivalent)

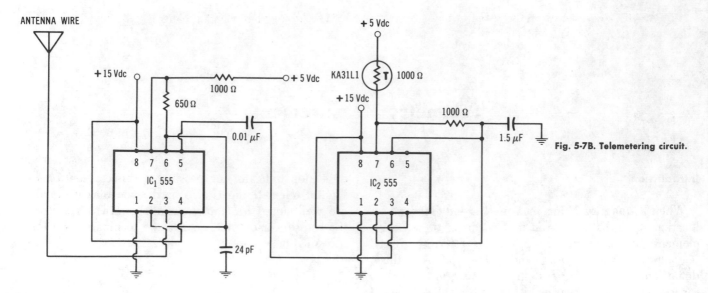

Fig. 5-7B. Telemetering circuit.

Am radio receiver
Wire: 5 feet (for antenna)
Connecting wires

Procedure

1. Construct the simple telemetering circuit shown in Fig. 5-7B.
2. Using the oscilloscope, record the waveform produced at the antenna of the circuit.

3. Using the oscilloscope, determine the frequency of the waveform produced at the antenna.

   Frequency = _____ Hz
4. Record the waveform produced at the output of IC_2 (pin 3).

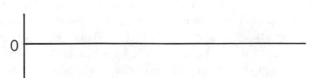

5. Determine the frequency of the waveform produced at the output of IC_2.

   Frequency = _____ Hz
   NOTE: The frequency as measured at the antenna must be between 550 kHz and 1650 kHz.
6. Obtain an am radio receiver and place it about 5 to 10 feet from the antenna. Adjust the tuner of the radio until the tone produced by the circuit becomes audible.

NOTE: The tuner of the radio should be adjusted to the frequency measured in Step 3.
7. Change the temperature of the thermistor connected to IC_2 by grasping it with your fingers. Observe the result of the temperature change.
8. This concludes the experiment.

Analysis

1. Write a brief analysis of this telemetry system.

_____

2. What are some possible applications of the principle of telemetering illustrated in this activity as an instrumentation method?

_____

3. How do the frequencies of the waveforms of Steps 2 and 4 compare? Why?

_____

_____

_____

_____

_____

4. In Step 7, what effect did an increase in temperature of the thermistor cause?

_____

_____

_____

_____

_____

pH Measurements

Introduction

The quantity of acids (acidity) and alkalies (alkalinity) found in certain liquids is of major importance in many manufacturing processes and must be carefully controlled. In order that the acidity or alkalinity of a liquid be controlled, it must be measured. The pH scale illustrated in Fig. 5-8A is used for measuring the acidity and alkalinity of liquids relative to pure or distilled water.

Acidity and alkalinity measurements depend on the separation of certain chemicals, when they are mixed with water, and the formation of both positive (+) and negative (−) ions. The formation of these ions alter the electrical conductivity of water, thus enabling pH to be measured electrically.

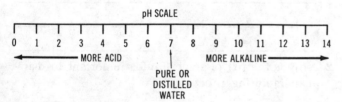

Fig. 5-8A. The pH scale.

Objective

In this activity you will examine the very basic principles involved with electrical pH measurements.

Equipment

Digital vom
Resistors: 10 kΩ (2)
Decade resistance box
Metal foils, 1 inch × 1 inch or 2½ × 2½ cm (2)
Styrofoam cup
Soldering iron and solder
Distilled water (½ pt)
Dc power supply
Table salt
Connecting wires
Spst switch

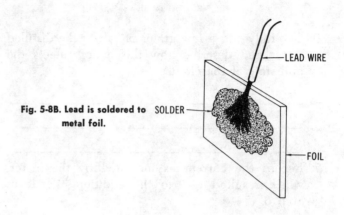

Fig. 5-8B. Lead is soldered to metal foil.

Procedure

1. Connect lead wires to each of the pieces of metal foil as illustrated in Fig. 5-8B.
2. Construct the circuit of Fig. 5-8C using the two pieces of foil with leads as the conductive cell.
3. Place a small amount of distilled water in one of the styrofoam cups. Insert the conductive cell into the cup as illustrated in Fig. 5-8E.
4. Close the spst switch and adjust the decade resistance box until the null state is indicated by the milliammeter.
5. Record the resistance of the decade box that is required to bring about the null state of the circuit.

$R = _____ \ \Omega$

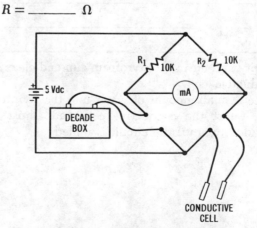

Fig. 5-8C. Measurement circuit setup.

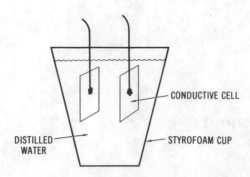

CONDUCTIVE CELL

DISTILLED WATER

STYROFOAM CUP

Fig. 5-8D. Test solution setup.

6. Carefully add a few grains of salt to the distilled water and describe how this action affects the null state of the circuit.

_____

_____

_____

7. Record the current as indicated by the meter, which resulted due to the addition of salt in Step 6.

$I =$ _____ mA

8. Add more salt to the distilled water and describe the effect as indicated by the meter.

_____

_____

_____

9. Record the current, as indicated by the meter, that resulted when additional salt was added to the distilled water.

$I =$ _____ mA

10. How do the currents recorded in Steps 7 and 9 compare?

_____

_____

_____

11. Empty and clean the styrofoam cup and clean the conductive cell.

12. Place a small amount of water into the styrofoam cup, insert the conductive cell, and adjust the decade box until the circuit is nulled.

13. Add three different substances to the distilled water and record the current as indicated by the meter. (NOTE: The substances added to the water should be added in equal amounts and should be common substances such as some soft drink, coffee, baking soda, alka seltzer, etc. The cup and conductive cell should be cleaned before the testing of each material.)

Substance	Meter Reading (mA)
1.	
2.	
3.	

14. How do the meter readings caused by the three substances compare?

_____

_____

_____

Analysis

1. What is meant by the pH of a liquid?

_____

_____

2. Why is the pH of some liquids important to many manufacturing processes?

_____

_____

_____

3. Why does the acidity or alkalinity of a liquid alter its conductivity?

_____

_____

_____

_____

Electrochemical Measurements

Introduction

The electrolytic cell (E-cell), along with the appropriate electrical circuitry, is a device capable of measuring operating time, cycle time, and general usage time, as well as average temperature, strain, or light.

The operation of the electrolytic cell depends upon the formation of "free" electrons and positive ions in an electrolyte. This formation of electrons and ions is brought about by a dc voltage source connected to the electrodes of the cell.

The E-cell is constructed with a "working" electrode and a "reservoir" electrode (Fig. 5-9A). The working electrode is gold while the reservoir electrode is silver. When the E-cell is connected across a dc voltage source (working electrode negative, reservoir electrode positive) electrons move from the working to the reservoir electrode. As this happens, positive ions are formed and move from the reservoir to the working electrode and are deposited or "plated" upon the working electrode. This action is known as "charging" the cell (Fig. 5-9B) and will continue as long as there is enough silver available to form the ions necessary to maintain the operation of the cell.

The dc voltage source may be reversed, thus causing the cell to reverse its action. When the polarities of the electrodes of a charged E-cell are reversed (working electrode positive, reservoir electrode negative) electrons move from the reservoir to the working electrode while ions move from the working electrode to the reservoir electrode. This action is known as "discharging" the cell (Fig. 5-9C) and will continue as long as there is silver "plated" upon the gold or working electrode. When all of the silver deposited upon the working electrode during the "charging" of the cell is removed through "discharge" action, current stops.

The charge and discharge characteristics of the E-cell enable it to be used as a very accurate measuring device. The amount of silver deposited upon the working electrode during its charging cycle is controlled by the charging current of the cell and the amount of time the charging current is allowed to flow. When the cell is discharged by using the same amount of discharge current as used during the charging cycle, the discharge *time* will equal the charge time.

A series resistive device can be connected to the E-cell to control its charging current. If this resistive device is a thermistor or an rtd, then temperature becomes the variable which controls the amount of E-cell charging current. With the charging time known and maintained during discharge, the amount of discharge current is indicative of the average temperature of the thermistor or rtd.

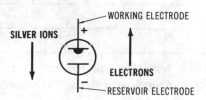

Fig. 5-9C. Discharging the E-cell.

Objective

In this activity you will examine the characteristics of a common electrolytic cell. *You must, at all times, maintain a charge and discharge current in the low microampere range. Likewise, the charge and discharge voltage must be maintained at a low level: 1 volt or less.*

Equipment

Voltmeter
Digital vom
Electrochemical cell: Plessey 560, no preset charge
Resistor: 100 kΩ

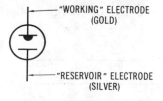

Fig. 5-9A. Schematic symbol of E-cell.

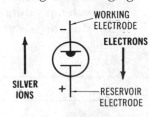

Fig. 5-9B. Action of charging the E-cell.

Spst switch
Watch
Dc power supply
Connecting wires

Procedure

1. Construct the circuit in Fig. 5-9D.
2. Close the spst switch and allow the E-cell to charge for 30 seconds. Record the charging current and the voltage across the E-cell in the space below.

   $I = $ _____ μA

   $V = $ _____ V

3. After 30 seconds, open the spst switch and reverse the polarities of the voltage source and the meters.
4. Close the spst switch and allow the E-cell to discharge. Record the discharge current and voltage across the E-cell during the discharge cycle.

   $I = $ _____ μA

   $V = $ _____ V

5. Allow the E-cell to completely discharge (near zero discharge current) and record the voltage across the E-cell.

   $V = $ _____ V

6. Describe how the charge current in Step 2 compares with the discharge current in Step 4.

   _____

   _____

7. Describe how the voltage across the E-cell in Step 2 compares with the voltage in Step 4.

   _____

   _____

   _____

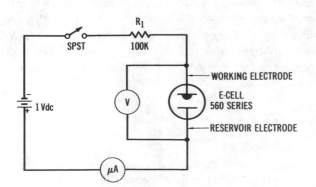

Fig. 5-9D. Electrochemical timing circuit.

8. Using the horizontal and vertical axis illustrated in Fig. 5-9E, draw a general graph that represents the voltage across the E-cell as the cell discharges.
9. Alter the circuit to the one illustrated in Fig. 5-9D.
10. Prepare to time the charging action of the E-cell carefully.
11. Close the spst switch and allow the E-cell to charge for *exactly* 1 minute. At the end of *exactly* 1 minute, open the spst switch.
12. Reverse the polarities of the power supply and meters in the circuit and prepare to time the discharge cycle of the cell.
13. Close the spst switch and record the amount of time required to discharge the E-cell (timed until 0 μA of discharge current flows).

    Time = _____ seconds
14. How did the charge time of the cell compare with its discharge time?

    _____

    _____

    _____

15. Alter the circuit as necessary to cause the E-cell to charge to the times as specified in the following table. After each charging cycle, discharge the E-cell and record its discharge time in the appropriate column.

Charge Time	Discharge Time
0.5 minute	
1.25 minutes	
1.5 minutes	
2 minutes	
2.5 minutes	
3 minutes	

16. If a thermistor of an appropriate value were connected in series with a charging E-cell, its temperature, and thus its resistance, would control the charging current of the E-cell. How could the average temperature of the environment of the thermistor be determined by discharging the E-cell?

    _____

    _____

    _____

    _____

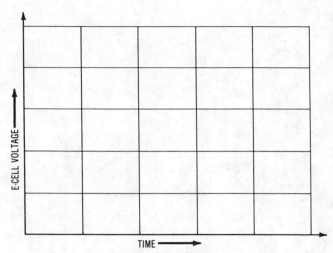

Fig. 5-9E. Graph of E-cell discharge.

17. If the thermistor were replaced with a strain gage, and the E-cell were allowed to charge and discharge, how could the average pressure upon the gage be determined?

18. The charge placed upon an E-cell is the product of current in microamperes and time in hours ($I \times T$) and is measured in microampere hours (μA-hours). If a charge of 10 μA-hours were placed upon the E-cell, how much time would be required to discharge the cell if the discharge current were 1 μA?

Discharge time = _____ hours

Analysis

1. How does an E-cell charge?

2. How does an E-cell discharge?

3. What is the difference between the working and reservoir electrodes of an E-cell?

4. How could an E-cell be used to determine the operation time of a machine over a period of one week?

5. How could an E-cell, along with the appropriate electrical circuit, determine the average temperature of an environment over a period of one day?

6. How could an E-cell be used to determine the average light intensity of an environment over a certain period?

NOTES

NOTES

NOTES